IN KILTUMPER

IN KILTUMPER

A Year in an Irish Garden

NIALL WILLIAMS &
CHRISTINE BREEN

BLOOMSBURY PUBLISHING
LONDON · OXFORD · NEW YORK · NEW DELHI · SYDNEY

BLOOMSBURY PUBLISHING
Bloomsbury Publishing Plc
50 Bedford Square, London, WC1B 3DP, UK
29 Earlsfort Terrace, Dublin 2, Ireland

BLOOMSBURY, BLOOMSBURY PUBLISHING and the Diana logo are trademarks of
Bloomsbury Publishing Plc

First published in Great Britain 2021

Design concept by Deirdre Williams
Design by Phillip Beresford

A catalogue record for this book is available from the British Library

ISBN: HB: 978-1-5266-3265-4; EBOOK: 978-1-5266-3266-1; EPDF: 978-1-5266-4557-9

2 4 6 8 10 9 7 5 3 1

Typeset by Newgen KnowledgeWorks Pvt. Ltd., Chennai, India
Printed and bound in Great Britain by CPI Group (UK) Ltd, Croydon CR0 4YY

To find out more about our authors and books visit www.bloomsbury.com
and sign up for our newsletters

To our children, Deirdre and Joseph

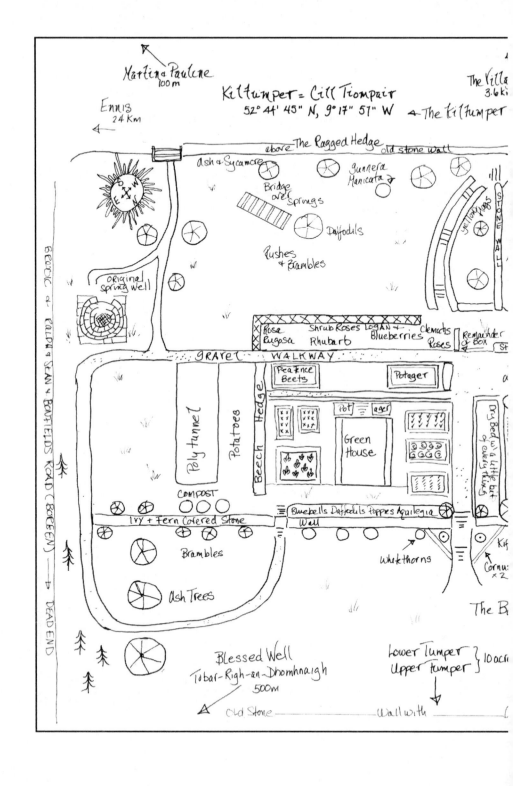

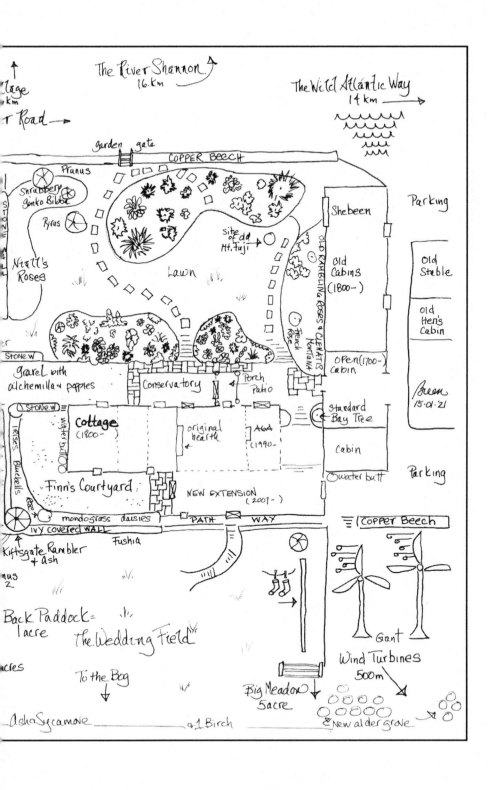

The River Shannon 16 km

The Wild Atlantic Way 14 km

Village Km

r Road →

garden gate

COPPER BEECH

Prunus

Shrubbery
Ginko Biloba

Pyrus

Niall's Roses

site of old Mt Fuji

Lawn

Shebeen

Parking

Old Cabins (1800–)

OLD RAMBLING ROSES & CLEMATIS MONTANA

French Rose

Open (1700–) Cabin

Old Stable

Old Hen's Cabin

Breen 15.01.21

STONE W.

Gravel with alchemilla & poppies

Conservatory

Porch Patio

STONE W.

Cottage (1800–)

original hearth

AGA (1990–)

Standard Bay Tree

Cabin

water butt

Parking

water butt

ROSES   Bluebells

Finn's Courtyard

NEW EXTENSION (2007–)

mondograss   daisies

IVY COVERED WALL

PATH   WAY

COPPER BEECH

Fushia

Kiftsgate Rambler + ash

nus 2

Back Paddock = 1 acre

The Wedding Field

Gant Wind Turbines 500m

To the Bog

Big Meadow 5 acre

acres

1 Birch

New alder grove

Asha Sycamore

The first lesson we might learn is that the point of a garden
is to be wonderful.

Henry Mitchell, *The Essential Earthman*

The way we are living,
Timorous or bold,
Will have been our life.

Seamus Heaney, 'Elegy'

This is a book written by two people, a man and a woman, who have lived together in one rural place for thirty-four years.

It is a book that has come as a natural consequence of a decision Chris and I made all those years ago, to give up the jobs we held in New York and move to the west of Ireland to try and make a life out of writing and gardening. We were young then, a leap was not daunting, and although sometimes mystifying to us now, the decision then was lightly taken. Like most people I suppose, we knew very little of what exactly we wanted to do with our lives. We were only following a prompting in our spirits that we wanted to live true to our own natures, whatever they turned out to be. It was a purely romantic impulse, equal parts foolish and rapturous, and it turned out to be one of the defining moments of our lives. Back then, there was no thought given to whether or not we had any talent, how we would actually make a living, nor what it would really mean to try and live from words and earth in a rural place on the edge of Europe in the last part of the twentieth century.

That place was, and, as of today, continues to be, Kiltumper, in west Clare. It is where Chris's grandfather was born, and his grandfather too, and as far back as can be known. And so, for thirty-four years, here we have been, raising two children, writing books, separately and together, in joy and hardship both, Chris painting and drawing, and together, both of us making a garden.

All gardens, as Henry Mitchell says, are wonderful, and wonder is not quantifiable. Ours is not a show garden, not the largest, or most anything, but it is wonderful, yes, and one which is, for want of a better way of saying it, us. So much so, that it is hard now to imagine our lives separate from it, and easy to believe that in some very real and meaningful way a garden becomes one with its gardener, and vice versa. Places not only become marked by people, but people by places too. Landmarked, and spirit-marked too, the relation is mutual and essential, because born of love. Many of these pages will be trying to attest to just this.

To both of us, for reasons both general and personal, this has come to seem more urgent and necessary in the past year. We are writing at a time when it is tempting to despair of the state of the world. There is a deepening gloom that the planet itself is in peril, that this may be the last century of life on earth alone, that by the next one another place, another planet may be needed, because this one will be in the throes of a man-made climate apocalypse which will be past the point of rescue.

So, it has occurred to us that the best way we can deal with this gloom we sometimes feel for the world beyond the hedge line, the Earth with a capital E, is to focus on the one with the small e, namely the piece of earth that we are fortunate enough to be charged with tending.

In this way, that tending and nourishing and growing of the garden has come to seem the central part of our lives.

We say this, having a heightened awareness that this too may be coming to an end.

We are both in our sixties now, Chris is not yet out of her bowel cancer, and still in the throes of a daily injection because she is at 'high risk of spontaneous fracture'. Nothing can be taken for granted. Then, this year Kiltumper itself will be in the midst of turbulent change with the arrival of the turbines of the wind industry, with their subsequent impact on the nature of the landscape and those who are living in it. Put simply, in the year ahead, when two turbines are sited on the hill 500 metres behind us, with their constant flickering and noise, we are not sure we will be able to carry on living here.

Recognising all of this, we have become aware of wanting to acknowledge to ourselves something of our life here, the decades of our being in this house and garden together, trying to make this kind of life, what that has meant, and continues to mean. What it means to try and live as writers, and gardeners, in this place and in this time in the world. Our children are grown and living in New York City. We know they carry with them this place we have made.

So, as we approached the cusp of a new year, we resolved to record twelve months of ordinary life just inside our hedge line.

The following pages are drawn first from both our journals, often brief notes jotted down at a small table at the top of the garden on a break between 'jobs', or written at the end of a day, or after a few days if we have been too caught up in the urgencies of spring or the abundance of summer. As in the gardening, and as is perhaps inevitable when two people have been married for nearly forty years – both to each other and to a single way of life – each of us have had a hand in the other's 'work', commenting, editing, addressing and readdressing, much in the way we do the garden. Each page then belongs to both of us and to this place, from which they are inseparable. As a result, it is our hope that, above all else, this will be an expression of love, green-fingered and white-fingered, of gardening and writing, of two people in one place, trying to grow and to make something beautiful, and that it captures a glimpse of the way we are living here in Kiltumper.

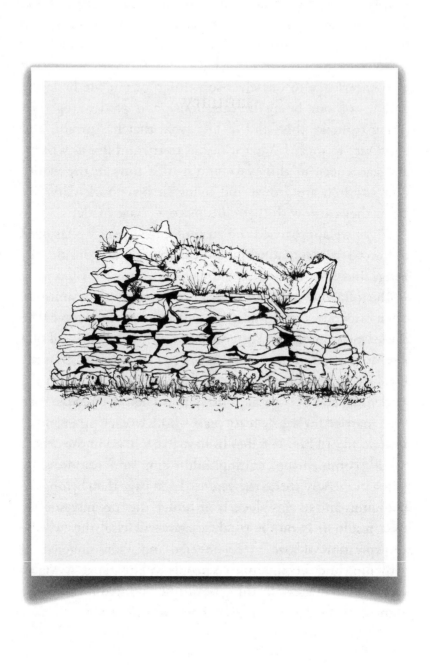

I

January

Ten seconds before the end of the year. The night is still. There is no rain. Standing by the old hearth, now fitted with a solid fuel burning stove, with two friends, Martin, and Chris's sister Deirdre, we have waited and watched time tick away the last minute of the year. As always, I am here and not here. Some part of me is thinking of the year ahead, not with dread exactly, but not with a full heart either.

This year ahead, I think. The shoestring and tightrope life we have been living here for more than half our lives is no more secure than it ever has been, there is a new novel to be published, yes, but we have no idea how it will be received, there are a half-dozen other writing projects not moving, including Chris's new novel, the roof of the conservatory is leaking, and this year the wind turbines will finally be coming to the farm next door.

By the time there are only seconds left in the year, we all take each other's hands, the way you might before a leap. We are momentarily hushed, and there is that particular inbreath I can imagine the entire planet taking in its great old age before crossing the threshold of another birthday.

'...five, four, three, two, Happy New Year!'

On the first stroke of midnight, following a tradition in Chris's grandmother's family, we clasp each other's hands and run out the back door of the house, taking the old year with us. Good riddance to it. Let it off. In a chain, like children, we run. To the stars and the spirits, to ghosts past, present and future, to the bare sycamores on the western wall, we shout wild joyous 'Happy New Year!' as often and as loud as we can, rushing along the crisp-frosted gravel round the back, turning left and left again to come in through the open stone cabin into the front garden – Happy New Year to it and all who have their homes there – and then, not delaying because we must get in before the last bells, we hurry to the front door to bring the new year in with us.

In this, we are aware we go counter to the tradition here, whereby it is bad luck not to go out the same door you came in, a custom probably born from the idea of the two great doorways of birth and death, and that you can postpone the approach of death by going back out the way you came in, or well, at least not prompt it. But in the fairy cusp-time of New Year, between the strokes, there is dispensation. You are in no-time and all-time, it seems to me, and can get away with it. And because my imagination is the kind it is, in those instants between one year and the next, in this hinge-time, which is neither here nor there, when time past and time future are present in the same moment, as we run in human chain past the stars to the front door, the thing that comes to me is *this place*, just that, this place that at this moment tonight seems, well, magical.

And I say that fully aware of all the times it is not.

All the times when it feels that our life here is a struggle, precarious and lonely, a committed engagement that takes all your wits and resolve, to live with purpose in a way that has evolved from both of our natures, so that it seems in some way organic, and true to ourselves.

This place. This place is the townland of Kiltumper in the west of Clare. It is comprised of less than a dozen houses scattered on two levels, Upper and Lower Kiltumper, in a lumpy hillside north of the village of Kilmihil. It is sixteen kilometres or so from the Atlantic at Doonbeg.

But, taking my cue from Wendell Berry's phrase, 'Think little', when I think *This place* I am thinking of just the portion of Kiltumper that is within our bounds, the piece of ground inside our stone walls, the long farmhouse and the line of stone cabins along the west that shape and protect what seems to be a secret garden as you enter it through what used to be the cart house, and, before that, was the one-room house where Chris's great-great-great-great-grandparents lived in the mid eighteenth century. It is a garden we have been making for thirty-four years – just as it has been making us. And it is this relationship, between us and the ground we have been living on, and with, that seems compelling to me now. It is a relationship that goes both ways. And maybe now that I am past sixty and have a different sense of perspective, I am more aware that the place has shaped us, physically and spiritually, in becoming who we are. To borrow and turn about Robert Macfarlane's idea of landmarks, this land has made its mark on us, just as much as, if not more than, we on it. We are certainly landmarked by Kiltumper, as I expect these pages will show. And at this time in the world, it strikes me that that seems a thing worth noting, the complementary and interdependent nature of the thing, the simplest story, of two people and their place in the earth, and the way they have been living together.

I have an added sense of urgency in this, because, as the year turns, I am newly aware of the fragility of our life here. We are older, we are four years into Chris's recovery from bowel cancer, and the coming of the turbines may make our continued life here impossible because of their proximity to our home.

So, *to capture something of the way we are living here, before it is over. To honour and celebrate it. And to put down something of the joy.*

In the no-time and all-time then, as we run out the back door and around the yard and in through the open cabin, before the last bell rings on the new year, before I pull open the front door to let us back inside, I am resolved that this year I will try. To show this place at this time, and our rich and precarious life in it.

We are our own first guests, and come inside to toasts and embraces and the bestowed hope that comes just from turning the page. Ahead of us, the year is still innocent and full of promise. The new novel is called *This is Happiness*, a risky title that Chris chose after she read that line in the manuscript, and I am resolved to try and remember the truth of that phrase each day we are alive and living on here in this adventure of an imagined life.

'This, is happiness,' I say to Chris.

She looks at me. Perhaps, having fallen more often, she is slower to take leaps of faith.

'This is happiness,' she says.

May this be a good one. Martin and Deirdre raise a toast. 'Here's to happiness!'

In the morning I walk outside onto the apron of the year. There is a delicious green stillness. I may as well be the only human on earth. It is not a far reach to imagine the earth taking its breath and taking its time, letting things be. But the nothing that is happening is of course a lie, and when I sit in a wool hat and jumper with a mug of tea outside by the front of the house, I see the birds have nearly emptied the feeder at the Japanese cherry tree. It takes only a few moments to sense the spring-in-waiting under the ground, the unseen engines of the earth thrumming, only a few moments to sense that all that will rise and burst and bloom in this place in the months ahead is now doing the business of gathering its energy in the dark and dreaming of the light. I know that sounds fanciful; it is and it isn't. There's more to Kiltumper than meets the eye.

I love this time. It is serene as an unopened parcel. Though the garden beds are mostly bare or showing only what, in my poverty of names, are X and Y, it has a winter beauty all its own. In January, the unopened garden makes dream-gardeners of us all. We can imagine what will come, and it will be wonderful, while at the same time, it will be *more* than we can possibly imagine. You can't out-imagine the reality of spring. It is greater than human dreaming, which is a good thing to remember, I tell myself.

~

The 'X and Y' of January. I love it. N's not great with names of plants.

The white flowers of hellebores peek through their browning, five-fingered leaves. A rosebud opens tentatively, maybe wondering why its neighbours have not done the same. Maybe it takes a look around and says, 'Cut me, please. Release me from this cold and bring me indoors.' The yet uncut stately spires of rusted astilbe plumes and the domed tops of sedums, the brown globes of echinops and the piercing needled rods of quaking oat grass, now half-bent by winter wind, hold on. All hold as if they carry the summer past for me to remember how it will be again. How valiant they are giving the garden that look that Piet Oudolf, the Dutch garden designer, achieves in his winter landscaping – leaving the leaves and flowers to die naturally. Further, in the lower bed, is the wine-coloured New Zealand flax, which sits like a giant pompom atop a woollen hat. Not N's favourite, but I have nearly convinced him it gives the garden colour in winter and sends your eye over the leatherleaf mahonia to the rose-brown leaves of the copper beech hedge. Each year we add another evergreen shrub to the framework into which the perennials will soon return. Refining and redefining.

My resolution is to try and be more grounded this year. I am a dreaming man, and will always be, but, at sixty, I have come to think balance important.

Don't be away in your mind so much, I counsel, and don't leave so much in the garden to Chris. In the way we are living here a pattern of work has emerged. Ours is not what my father might have recognised as 'a working life', but comes under the general classification of 'alternative', with all the hardships and rewards that come from going your own way. Usually, I try and write in the mornings, and try to work outside in the afternoons. Chris works outside much of the day, and would be there longer if light allowed. The garden is where she is most at home; it is the place where she forgets time, forgets that she is trying to make her way past bowel cancer, forgets pain I suppose; it is where she is free.

But, too often, I have gotten lost in words, the morning spent writing has bled into the afternoon, I have postponed the urgencies of the outside, and left too much to her in the past. This year, I am resolved to try and remake the balance, and be in the garden more.

Which is easier said than done. Not least because time spent outside brings no income, and a writing life is a blind one, full of white pages and no certainty of any future. But here's the thing: it feels *natural* to be outside in this place. Maybe not more so than any other country place, I can't speak to that, only that the given beauty of your surroundings, the sense you have here of being not close to but *inside* a green landscape, means that it feels in some elemental way right to be outside in it. I'll come back to this again and again this year, I know.

What I should be doing with my life is a question that entered my mind probably at eighteen, and it hasn't left. One early answer was to try to become a writer, and then a better writer, and keep trying to write one good book before I die. But as I have grown older, the central plank of that has shifted somewhat. Is it enough to just write? In this time of the world especially, it seems not. At least not when you are fortunate enough to be living in a landscape like this. Chris and I have spent more than half our lives here, trying to make this place more productive and more beautiful, and the evidence of that is everywhere. But on this first day of this year, sitting in the garden in Kiltumper watching the birds of winter feed, I feel I need to renew my commitment to it, and do better. 'Be with you half of every day,' I say to the garden, knowing it is a promise I will fail, but keep the intention in sight anyway.

I watch the birds feed and fly away a short distance to the edge of the cabin roof, and then come back again, and I try to figure out if there is a sequence, if they know which bird goes next. I spend a good twenty minutes at this, and just when I think I have worked out the order one bird confounds it, and then a magpie whirls down and startles them all, and a new sequence will have to be evolved. I leave them to it, but resolve that in the year ahead I will try and be a better noticer. Notice something each day. Just that will be a worthwhile way to be living, I decide. A moment later, I realise that noticing anew each day won't be so easy either. It will require conscious effort, at least at first. But then, should you be so conscious when you are outside in nature? Is that natural? How will I be at ease and alert at the same time? 'We'll see' is the conclusion of that debate, and takes the way of all humanity by letting me off the hook.

One day, around the time of the full moon, the Wolf Moon, the Hungry Moon, so called by the Native Americans because the wolves were hungry in January, the sky turns blue after days of wicked rain. It, like other blue days in other Januarys, spreads the promise of spring above us, albeit briefly, but who cares as today I can see only blue and as I have written before elsewhere, for me blue is the colour of hope.

A few years ago, I hung a bird feeder, filled with seeds, on a branch of the cherry tree. Ever since, the birds have delighted, as do we watching them from our seats at the table in the room we call the front kitchen. When N is writing in the conservatory I usually sit here to write. Between the wire cage of the feeder and the twisted and gnarled branches, clothed in lichen and green puffs of moss, the limbs of the Japanese Cherry tree, Mt Fuji – now dead I am sure of it, although Himself doesn't agree – he's in denial – half a dozen birds feast. Blue tits and coal tits, robins, goldfinches and greenfinches and little brown wrens. I am waiting for the return of the crazy bird, who last year seemed more preoccupied with our large glass doors than with the golden nuts and seeds. Thump. Tap. Peck. Tap. The little brown tit flew above the cat then and headed straight into the window. His light weight negated the impact against the hard window and didn't faze him one bit. He kept coming back for me. It was as if he simply feather-kissed the window and sent a message. Of love? Of keep up the good work? Of hello we are all connected? Don't worry be happy, every little thing's gonna be all right? Himself was writing in his journal, I remember, but with the incessant tapping of the bird's beak he says his thoughts were not merely being punctuated but speared through.

What was he doing? I decided the bird was up to something profound. Trying to tap words into Himself's imagination. Once the message was received, however unconscious, the bird desisted.

Birds are a lot smarter than we give them credit for. Research tells us that 75 per cent of their grey matter is made up of a sophisticated information processing system that works much the same way as our own cerebral cortex. Although their brains might be small, scientists conclude that bird brains they are not. (Betty the crow could fashion a hook out of wire to access food. Alex the grey parrot could speak a hundred words with meaning!) And with the beyond-dismaying news now that bird numbers are diminishing everywhere, perhaps their message is simply Stop destroying our world or we're leaving. Write about that.

Something about where we are: right-angled to the line of stone cabins that run down the western side, the house is a long low farmhouse typical of those built around the beginning of the nineteenth century. Its walls are stone, three feet thick. When we first arrived here, on 1 April 1985, it had four rooms, each one the depth of the house, so that the front door and the back door were in the same room, where the floor sloped twice, both down and out, so that, with an intelligence born from experience, a bucket of water tipped at the back door would run out the front one or into the grate of the fire. A built-in pine dresser of indeterminate age faces a ten-foot-wide hearth, which, for our first thirty years here, had a grate on the floor and a chimney you could look up into and see the sky. You could climb up and out of it if you wanted, and down it; Santa could bring a bicycle. Of course, down it too came the black hailstones, bouncing out onto the floor so you learned to raise your feet during the fiercest moments of a downpour. A first lesson of the place maybe. The second to have a towel handy. For thirty years we lit the fire on the floor most days, that rising column of pale smoke one of the signs of life of the place, and for Chris a continuity with all those whose lives had literally begun next to, and who had lived about, the same fireplace. Stretching across above it is a wooden lintel, probably bog deal, which we only discovered when we put the stove in. It had been here since the day the house was built. It is not exactly carpentered, but clearly handmade, and scored in a hundred places like some ancient totem with the marks of adze and chisel.

This was the place of the family who were once known locally as The Long Breens (a taunt from history to Chris, for although her father and brothers in America are all over six feet, she is five foot four, almost). When we first arrived, from the older people, we heard some vestiges of the Long Breens legend, of horses and apple trees mostly. Extraordinary what survives of us. A great-grandfather called Jack-of-the-Grove, in whose name alone was unpacked story. Over the decades we have adapted to the house and it to us; we have added rooms, mostly with the kind of short-term and necessarily near-sighted planning of people without a steady income. Several of the additions can be traced to books. A book came out, a room was added, or fixed up. So, the place grew organically out of the books written in it. And in that way came to look like the life that was lived there. That's how I've come to think of it anyway. It is odd, and individual and quirky, but in a way that has come to seem important to me, this place is us. And somehow, it all works. After a fashion.

In the eighteenth century the Long Breens lived in the one room of what we call the Open Cabin, a stone building with a high roof and a ledge for bedding halfway up the wall across from the fireplace. It is unknown how many Breens exactly lived there or for how long. But by the start of the nineteenth century the place had taught them an important local lesson: don't build your house facing the west wind. The new house they built faced south, and, crucially, they didn't knock down the old cabin or take the stones. They left it there as a cow house and cart house, and we are grateful to them for that. For the shelter of the old cabins is what allows us to have the garden.

'Could you describe the garden?' he says to me.

I know it too well to describe it, I think. But here goes:

Facing south and sloping down from the house, the front garden is about thirty metres wide by eighteen metres long. As N has written, it is bordered all down the western side by stone-walled cabins, and along the southern end by a mixed-shrub hedge that runs above and alongside the road below it. The hedge has a small crooked old iron gate in the centre.

When we first arrived in April, thirty-four years ago, and spring was already upon the garden, I asked Mary Breen, who *then lived down the road, but who had previously lived here, about the little shoots that looked like red asparagus appearing along the front wall of the cabins still partly whitewashed with lime. Could they be wild asparagus?*

Mary didn't know. (She didn't know what asparagus was.) She said the plant had pretty white flowers. Those red tips soon grew and grew and very quickly too.

My ancestors had planted Japanese knotweed! It would be a few years until we discovered the nuisance plant that it is, but after decades of pulling and pulling and pulling its population has decreased. It makes me think that maybe my great-grandmother Breen née Ryan was also a gardener and had taken the advice of William Robinson, the famous Irish gardener and writer who likely introduced Japanese knotweed to Victorian gardens, suggesting it was an exotic plant worth having. Advice that turned out to be calamitous.

Along the cabin wall are old-fashioned pink ramblers which climb all over the cabin roof and sometimes find themselves inside the cabin, having infiltrated between the stone and roof. That they have survived this long is a testament to the shelter offered by the cabins. This wall is also home to a giant Clematis montana. It extends nearly thirty feet, climbing up and covering most of the cabin roof. It is spectacularly and completely untamed.

The front garden had once been a working farmer's garden: potatoes and cabbages, onions and carrots and rhubarb and a lot of manure. It was a jungle of brambles when we arrived, having been neglected for five years. Now two large, irregular, nearly kidney-shaped perennial beds span the bottom of the garden, divided by a curving path to the gate. When we first started out there was a straight path from the front door right down to the gate, but since we have been gardening here straight lines have gradually begun to disappear. Not sure what that says about me, or us, or Kiltumper. Both of these beds have specimen trees nearby, one the flowering Japanese cherry, a pyrus, an autumn-flowering prunus and the tall skinny ginkgo.

At the top of the garden a long herbaceous bed borders the path along the front of the house. The bed is about twenty-three metres, divided by some flagstone steps. It is the main bed of the front garden, gets the most sun, and attention. Think poppies, delphiniums, lupins, helianthus, peonies, astilbe, rudbeckia, lemon balm, asters, agapanthus, scabious, baptisia, achillea and more.

On the eastern boundary of the front garden, framed by an old stone wall, which occasionally gives up an old boot or bit of crockery, is what has become N's rose bed with about fifteen rose bushes.

Then there's the back courtyard gravel garden, which is sunken, the house rising on two sides of it and an ivy-covered stone wall rising on another. It's our nearest thing to a walled garden, however only a small portion gets sunshine.

On the eastern side beyond the front garden is the field that has always been called the Grove. Its lower half is sloping and damp, with several springs in it and a giant gunnera. It has the polytunnel on the upper ground, and a number of very old sycamore and ash and two holly trees. It measures forty-six metres by thirty-seven metres and includes the raised vegetable beds and the greenhouse.

A strong green beech hedge separates the vegetable garden from the latest addition, a polytunnel, as 'the garden' continues to move east away from the house.

A shale-gravel walk, a hotbed for weeds, passes the polytunnel and ambles towards the easternmost aspect of the garden near the old spring well, and through a gap in yet another stone wall and into a very small grove of pine trees and more sycamores and ash where the path reappears in the Wedding Field and brings you back again to the glasshouse. I like a path that goes somewhere.

We haven't planted to the west where in spring the grass is full of daffodils and in summer montbretia (another invasive but pretty plant) because it will be in full view of the turbines. That was next on my list, another kitchen garden but not now. My idea of gardening is a bit like a garden weed with roots that just keep going like couch grass, or scutch grass as it is known here.

I'll have to stop making new gardens – unless, unless I extend into the five-acre field, called the Big Meadow, at the back of the house… No. No, I won't. Not now. That would bring me even closer to the wind turbines that are coming.

This morning I walk down the still-frozen pea-gravel path – how delightful is the small crisp crush of your boots on it – to the polytunnel. There is not a lot in it in January. Mostly we are concerned now with improving the soil there.

Let me say something here about soil. I am reluctant to admit that it has taken me this long to understand that Chris was right all these years when she was telling me about the importance of our improving all the soil in Kiltumper. I suppose I had been thinking that she was being, well, alarmist, my ignorance probably founded on the thought that things had been growing here for hundreds of years. There was soil and things grew in it, is, I suppose, a fair summary of the writer's thought, looking out at the garden through the window. Characteristically, I probably came to a different understanding first through the work of a writer, in this case the American naturalist, essayist, novelist and poet, Wendell Berry, who writes superbly and urgently about nature, and eloquently about the importance of soil. To say soil is the fundament of the earth seems a redundancy, better perhaps to say it is the only substance besides water that guarantees our continued existence on the planet. And all topsoil is continually being washed away. That's a hard thought to take on, but I have taken it on and with it the understanding that soil is something that is not just there underfoot, but that it is something that needs to be *renewed*. Somewhere Berry has written that a measure of a good life might be, at the end of your days, to be able say you had added an inch of topsoil to whatever ground you were living on.

Now, the idea of that has stayed with me. For years we have, with varying degrees of success, made two containers of our own compost. But while your heart lifts when you finally adjudge the dark wormy humus ready for use, it falls when you see just how little six months of composting has made. Chris rations out handfuls of the stuff. Only if we lived on a patch of ground ten feet square would we have added an inch. This year that'll be a focus, I resolve. Resolutions are crisp and fresh at New Year's, and for a few days at least infallible. Before the year is done, we will have improved our composting, and have harvested some seaweed from the shoreline. That's another promise. And as with all things for the writer, by writing that in my notebook, it takes on an actuality in language first. But this year, I add, it will also actually happen.

In the polytunnel, you can see first-hand how poor the ground is. The tunnel is only in its second year, and in its first did provide plenty of vegetables, so you can't be too hard on it. It also had the benefit of allowing us somewhere to be working when the heaviest rain was pouring down, and we needed to escape outside into physical work that we knew would produce something – an important factor for writers who are never sure the hours of their work will amount to anything, well, real. But in the tunnel, you come face-to-face with the actual earth of Kiltumper. Because of the confines of the space, because of the forensic light, and the absence of other green distractions, in there you are in a laboratory of sorts. It is artificial, but necessary to compensate for, I am tempted to say, the deficiencies, but am learning to say the eccentricities of the climate, and extending the growing season, so I have made peace with it. But, as I've said, in the tunnel, there is no escaping the reality of how poor is the soil. There is only an inch or so of topsoil and then you are into a dense pale grey-brown stuff that clumps and bonds together. You dig it up in lumps, revealing gley, waterlogged soil that doesn't let air or water through it; so, there are no worms. Rushes thrive in this environment. Oh joy! Chris said when we read up on it. A catastrophe in the soil department. Since we got the tunnel up, we have been committed to improving this. We have brought in barrow after barrow of rotted horse dung, we have added all of our compost, and often I have come out to find Chris in there, kneeling on the soil and going at the lumps with a small hammer. 'If we don't break them, nothing else will. Look, they're hard as rocks.'

She gets some satisfaction from the hammering. Let's say that.

The soil remains pretty poor. When you pour water onto it, it runs off. You might water until the ground is dark, but dig two inches down and the earth is dry as dust. Which is vexing. You want to be bringing life to it. That's somewhere in the DNA of all gardeners – in a hopeful version, in all human beings – and the failure smarts. So, in Kiltumper, the tunnel is its own test case, and since we are more in charge of what happens in there than anywhere else, since we control the climate and the soil, we will make it more productive than elsewhere. If you make an artificial environment like this, it's the least you can do to make it thrive, is the way I think of it. The polytunnel is not for beauty, it is for food. So, *come on now.*

The main focus in there right now is a dozen purple-sprouting broccoli. Having read in an old English gardening book that all brassicas enjoy a beer, in a fit of sentiment and dreaming, on Christmas morning I had quietly gone out and given each of the twelve plants a good drink of old Harp. I poured it in at the base of the thick stems and watched it gurgle sourly down, half expecting to see the plants keel over there and then. Was it a joke perhaps? Could it be true? I had come across it nowhere else, and no one else I mentioned it to had heard of it. Once the beer went down, I wondered if you were supposed to pour in some water as a chaser. Was one pint per plant too much or too little? The old gardening book hadn't said. A brassica likes a beer was all that was written. On Christmas morning I decided not to add any water, but when I left them and went back to the house, I really wasn't sure that I hadn't inadvertently killed the crop.

Well, in the week since Christmas, the plants hadn't changed a bit. They had been planted in September and at this stage were four feet tall. But there was no sign of any sprouting, purple or otherwise, happening. Instead, just giant green leaves that could be seen by Martin and his missus, Pauline, walking the road – '*What are ye growing up there?*'

Then, this morning when I went in to look at them, I could see the tiniest purple tips, just nubs really. But there they were. The sense was of a thing happening, and you were gifted this element not only of witness but of participation.

Never mind how they would taste, never mind that you would have the only fresh purple-sprouting broccoli hereabouts, that it was not something you could get in any shop, except at the farmers' market thirty minutes away, the thing that was maybe more profoundly nourishing was the fact of your being, well, connected to it. And that made my spirit smile.

I came in to Chris and with a schoolboy pride announced: 'The beer worked. They're sprouting.'

She just looked at me. 'Yes,' she said, 'along with the staking, and the watering, and the kneeling down searching of the underside of their leaves, and the picking off and collecting of insects, the beer worked.'

'Right.'

*

I'll admit that I have had to come to terms with the word 'garden'. I can't trace why exactly, but both noun and verb had weak associations for me, suggestive of hobby, retirement, a kind of gentle pastime. Nothing necessary or urgent. The phrase 'we spent the day in the garden' conjuring indolence, something indulgent even, set aside from the business of real life. Now, some of this comes from my upbringing in Dublin where the garden was small, and the jurisdiction mostly of my gentle grandfather who visited once a week and tended the roses. Some of it from the standard portrayal of gardeners in books and film. And some too from the Victorian notion of gardening as containment and control, the anti-wild impulse, the categorising, boxing, listing, shaping that the novelist John Fowles pointed out was essentially an attempt to reduce nature to human dimensions. An impulse that is not in my nature. But for years here, I think I thought that work in the garden was not 'real work'; I would do the real work first – which, ironically, was entirely imaginative – and afterwards go out into the garden and join Chris. I don't think this was out of some sense of the masculine, that gardening was not man's work, but it was certainly true that I regarded time in the garden as the less important part of the day. I don't any more. And I expect the aim of a good portion of these pages this year will be to acknowledge to myself just that.

It doesn't escape me, of course, that to do it I will be bringing the garden inside, into these written pages. But the truth is that over thirty-four years the two have become inseparable.

In thinking of this, I have been reminded of another writer-gardener, or gardener-writer, Chris and I met years ago. After he had written about us in his *Irish Times* column, 'Another Life', we drove to Mayo to meet Michael and Ethna Viney and to see their garden. The Vineys were already legendary, having left a life in journalism in Dublin to try and live a self-sufficient one under a mountain and beside the sea in west Mayo. Michael is a gifted naturalist, artist and gardener. Each week he wrote both about the challenges of Atlantic gardening as well as on broader environmental issues. He is an eagle-eyed observer of plants and animals and to the young man I was he seemed to have an encyclopaedic knowledge of the natural world, of which I was in awe. A soft-spoken man, Michael welcomed us, and brought us around their garden. The thing that was immediately striking about it was that, in order to shield against the salt winds, they had planted lines of fuchsia hedging around each bed, clipping these at three feet high, so that their garden was a maze of low hedge walls with vibrant beds inside them. These were garden 'rooms' long before that become a fashionable gardening concept. There was nothing fashionable about them in fact. They were not only practical, in that place, for what they wanted to grow, they were a necessity. From the nearby shoreline, Michael and Ethna frequently gathered seaweed and so literally made the soil from which their vegetables grew. And what they grew was bountiful. In visiting them in their place, I think I had my first sense of a couple inseparable from their landscape. They brought us in to a lunch entirely from their garden, and we spoke a little about our own dream of such a life in west Clare.

Now, this was over thirty years ago, and I cannot remember anything we said. What I remember is how impressive they were, focused, resolute, not in the slightest bit airy, but with a deep-rooted contentment, and how kind to two young people dreaming of their own version of another life. What I remember is the fuchsia hedges and the focus and energy in their making a life in that place, what it required of them, and how they had met that requirement. I was too young, and too inexperienced in gardens, to fully realise what this entailed. And I was probably still too attached to that inherited idea of gardening as a pastime.

But now, now that I am probably the same age Michael was then, now that Chris and I have spent more than half our lives trying to make a garden-life, and life-garden, I think of that visit, and am grateful for the inspiration it gave, for the idea that trying to live closer to nature was a valid and worthwhile one, and for the example of a different, vigorous and vital way of thinking of the word 'garden'.

What I have come to, I think, is in the origins of the word itself. In the guard that is in garden, in the sense of an enclosure, and, importantly, something protected. This, more than the idea of cultivation, is what seems central to me these days. In this enclosed space in Kiltumper, we will protect and nourish what is here.

That is what we are doing when we say we are gardening.

And there is plenty to do in January.

I think it is probably universal: on the first day back in the garden after Christmas, no matter how you counsel yourself not to try and do too much, no matter how you say 'take it handy', you do too much. Because everywhere you look there is work to be done. The days are mild and damp and already the tips of the daffodils are above ground. It is not yet Little Christmas, but the pulse of the spring that is still underground has entered you. You are the same as all that holds within it the returning life, and your version of that is to be outside moving above the beds, tidying, thinking to cut back all that you left to turn brown and black and amber, what you left both for its colour and architectural grace in death, the spires and seed heads, the stems and stalks that now, like Christmas ornaments, must go. You are limited only by the clockwork of mud. Too much in one place and your boots will make muck underfoot, so we both work in movement, like birds, now here a bit, now over there.

There is a special hush in the first days of January. Maybe it is just that the busyness and company of the Christmas has made the silence outside deeper, but I can hear Chris's secateurs deadheading across the far side of the garden.

✄

What is it about Japanese anemones that allows them to grow so well? As Alan Titchmarsh says, 'Once they are established they will romp away and are as reliable as the tide.' Put another way, they are exceedingly robust and easy to grow and can become invasive. But first, they are not Japanese at all, having originally been imported from China and then cultivated in Japan. Robust and abundant they are in our garden, but like many things I observe about my own successes or failures – they are in the wrong place or places, and it appears I've planted the wrong variety, too. They like to grow not where you've planted them but where they want to grow. Wind-scattered seeds take root wherever they want, which is why I think they are also called the Windflower. (Kiltumper could start a nursery of just Anemone hupehensis.) The pale pink colour variety we have (Anemone x hybrida 'September Charm') is neither elegant like the stately white 'Honorine Jobert' nor pretty as the deeper pink of the Pretty Lady series. I'll be on the lookout next year for 'Lady Diana' or 'Lady Julia' or 'Lady Susan'. I know a perfect place for the other ones, along the verge of the road that passes our house. In a few months the lovely 'ditch' with its ancient ivy-covered stone wall – home to ferns and perennials I've transplanted over the years in memory of Mary Breen – and old ash trees will be demolished. In any event, in this moment, with a few flowers still in bloom, which is a bonus when you need to bring something indoors to brighten up the guest room, they are ready for the chop because unlike the graceful dying of other plants the leaves of anemones blacken and look sickly and diseased like a plague. It's an on-your-knees job, your head lost in the centre of the dense thicket of them.

I leave the piles of cuttings on the grass for Himself.

~~✦~~

There is such hope in a day like this. The garden-to-be still wonderful, unimaginable even. If I stop and stand and look across it this January day – even though I have known it up close and personal for decades – even I cannot fully picture all that is to come. I might have a mind-picture of it in May, say, when the poppies and the lupins are in full display, but what is in bloom as they fade? The fact that it is constantly every day changing defies pictures, defies framing, and it is best not to try. In our early days, when visitors came in wintertime and looked down over the blackish brownish beds, somewhat apologetically I would try and talk the flowers into being. 'Over here in June, it is all…' But the garden has taught me a broader appreciation of itself, not just the glamour of its young days but all of its time. Which may be a natural function of age, I don't know. Enough to say, I love the place in January and leave it at that. In January, all gardeners have dream-gardens in their heads. And the chance to put right what went wrong last year. To move or divide what was discovered too big or in the wrong place last year. That this happens every year, that every year the dreamt garden will be better, can be improved just a bit, seems to me a good small hopeful thing to think of, these days especially. (My mind being the way it is, it is only a short step for me to think the plants, in first waking underground, are thinking the same: this year the gardener will be better.) And while, too taken by shape and colour, by the sheer vibrancy of the plants in spring and summer, my memory of exactly which got too big or were in the wrong place is poor, Chris remembers. With a forensic eye, she looks at a multi-prong of black stalks, just inches above the ground, and says 'You got too big for here last year.'

Is it a sign of a pure gardener, that she can throw no plant out?

As a result, each decision is a double one, we can move this yes, but where?

And so it is we go through the January day, and into the low light of the afternoon, quietly, from bed to bed, cutting back, or moving aside wet dead leaves and withered stalks, finding new spaces for divided plants. If none can be found there is the Blessed Well up the road, or the ditch outside.

A robin follows wherever we go. Despite all good counsels, we do too much. Driven by engines of hope and renewal, we exhaust ourselves. By four o'clock, the light is gone and a blanket of cold lies along the back of my shoulders. Inside my Christmas gloves, my fingers are frozen.

'I think I'm done for today,' I say.

'I'll just stay out for a bit longer.'

'It's almost dark. We've done a lot.'

'I know. I'll be in soon.'

I go inside, come back out in the fallen dark an hour later to call her in.

2

February

Chris goes out this morning to gather rushes. When she comes back in, she has a good bunch, and as she lays them on the table, I realise that she has not just gone and cut a clump at random, but rather she has selected each for its density, pliableness and especially length. That's how she is. She puts on her glasses and in the absolute quiet of the house this first day of February she makes Saint Brigid Crosses. I'm not sure even Brigid was as careful in measuring the lengths.

Niall wonders why I can't do this tomorrow; wait until tomorrow, he says. The weather will be better. Today is not blue, not warm, but damp. It's tradition to cut them on the eve of St Brigid's Day, not on St Brigid's Day, I tell him. The place where rushes grow is on wet ground, and trudging along, my wellies sink. If, as the tradition goes, the rush cross is meant to keep fire and hunger and even evil away from the house it's a small thing to get it right, don't you think?

After an hour or so, she has some made. She puts two in large envelopes to send to America, one to the Breens in California and one to our daughter in Brooklyn, and another she hangs on a nail in the front room.

Something resonant and moving in this. In the simplicity and ancientness of it. It gladdens some deep part of me, and though the day outside is wind-tossed and the rain comes in waves, the Brigid Cross says: *Believe, there is a spring coming.*

*

To leave the place better than we found it.

That's what I write in my notebook, then lean back thinking how small is that sentence and how immense the idea.

To fully inhabit the place you are living.

That's another one.

I have been reading Robert Macfarlane where he writes of reading the work of islomaniacs, Tim Robinson on Aran, and Lawrence Durrell on Corfu, and the belief that they might come to completely know a chosen place because of its boundedness. He says that, ultimately, they each found this impossible. I think I already understand that even our garden in Kiltumper has more to it than we can know in one lifetime. And this is not a drawback but an epiphany, a clarifying moment in which Chris and I are only specks on the eternity of the place.

Good specks, please God.

*

Today, I have been thinking again about the way we have been living our life.

It's not something I think of too often because I'm too busy living to think of Life, is an answer I might have given if asked why. In our thirty-four years there have been many books, from both and each of us, and these have seemed a natural expression and measure of our life in this place. But the measure I am thinking of today is the garden – or at this stage gardens – inside the ragged hedge that runs along the stone wall.

~~~

Ragged? What does he mean *by ragged? The stretch of copper beech, a recent addition to our garden hedge line, which we planted two winters ago, is not ragged. When we can afford to extend from the westernmost corner of the garden fully across to the eastern side, spanning about thirty metres, we will replace the whole of the 'ragged' hedge N refers to with copper beech. Honestly though it has a character all its own, but in summer is a nightmare for N to keep trimmed, which is one of the reasons we want to replace it. He's fallen off the ladder more than once with the electric hedge trimmer buzzing in his hands. (Worse, three times now he has cut clear through the electric cable. Himself is not really a man for machines.)*

My ancestors stuck bits of plants here and there on top of the stone wall that rises from the road. (We've also found old boots and blue medicine bottles and bits of pretty crockery.) In those days, having no money to spend on plants, they simply stuck cuttings from elsewhere into the ground. And they grew. There's wild rose mixed with privet and poor man's box (Lonicera nitida), *which needs three trimmings a summer, and of course fuchsia – the Irish name is* deora Dé, *which means tears of God, but you only see those tear drops if you let it grow and, dare I say it, the hedge, where it is, is in the wrong place so it is kept trimmed and we rarely see its flowers.*

~~~

For some time of each day, in all seasons, we work in the garden.

I have been thinking of the meaning of that, what it has meant and what it continues to mean to the quality of our living here.

I was thinking this while working outside in the ditch across from the front gate, rescuing a white fuchsia bush that had been given to us years ago by Paddy Cotter. Hearing that Chris was a keen gardener, Paddy had arrived one day with a grin and a cutting from his own garden. The red fuchsia was everywhere, but the white one, now the white one was special. A gardener like Chris would appreciate it, Paddy thought. And he was right too.

By the unfathomable logic of gardens, the place where the cutting grew best was outside on the side of the road across from the gate, and each year it bloomed there, pinkish white, pale and delicate, drawing small notice perhaps from those who passed. But we noticed. That's the thing. It was a thread in the fabric of the place, in its history was its meaning, and though Paddy passed on, the white fuchsia retained a redolence of that spirit of community that informed a man one day calling to your back door with a smile and a cutting from his garden. As Chris has said, it was from cuttings that all gardens here were made. They are borrowed gardens, all stitched together into the fabric of here, is how I think of it.

Now, as in so many places in mid-west Clare, in Kiltumper, wind turbines are on their way. In the near future they will be coming along the narrow road with several bends that passes the house, and to make way for them, for the one journey on the one day they will make, the road must be widened. The centuries-old stone walls and ditches on the southern side will be bulldozed and the margins expanded for the night-journey of the two turbines. The bends will be straightened. That this will change forever the character of the road is an argument that convinces few. Change is the only constant in life. And most people say: Won't you have a nice straight road now?

If we didn't rescue the white fuchsia, the digger would take it in a breath, along with the many perennial plants Chris has planted there, and, with it, it seems to me, Paddy's grin and the generosity of spirit we met so often in the old people when we first arrived here. That fragment of – I believe the appropriate word is history. That narrative of past events *particular to a place* – well, that would be gone.

So, as the occasional car or tractor passed this afternoon, I was teasing out the roots of the plant from where they had grown in and through the stones of the roadside. I worked diligently, and must have seemed an odd sight, man in the ditch, hands probing the mud for something in ordinary measure not that remarkable.

And that's what made me think about the way we are living.

By the time I had brought the bush in three segments inside the gate, and planted them in three different places in the hopes that one would survive, a couple of hours had passed. *And that's what I did with those two hours*, I thought. That's what I would have to answer if anyone asked what I did this afternoon. I rescued Paddy Cotter's white fuchsia.

And, somehow, I felt it mattered.

～✤～

It's a thing. Plants in wrong places. Here's another one. I've noticed that this year yet again I have forgotten to move the small crop of snowdrops growing beneath a myrtle bush. The myrtle bush is a Mediterranean plant and while I love it N wonders why I keep it where it is. One, because I don't know where else yet to put it, and, two, because I like it and don't want to risk losing it. He doesn't seem to appreciate the lovely shape of its leaves and he doesn't know that if we wait long enough and we get a Mediterranean summer (it happens) it might bloom with small white flowers. It wants to be in better soil, not one so full of clay, and it needs some protection from the cold wind. But it has done a nice job of sheltering the snowdrops. Will I dare to move them both?

Back to the snowdrops. An Irish expert on Galanthus, *Paul Smyth, says that you should grow snowdrops where they want to grow. Ideally in a woodland setting. Under the myrtle bush is perfect because the myrtle is under a tall copper beech tree. A woodland setting I should think. The great thing about snowdrops is they are very adaptable. We like that in the Kiltumper garden – a bit of shade but can take full sun. Like me and Irish weather, they can tolerate damp soil but don't want to get waterlogged.*

And to be quite honest I am tiring of the storms that each year seem to be coming closer together.

It's decided: I won't move either of them, but I'll divide the snowdrops and take some cuttings from the myrtle.

Now, to be clear, Chris is the real gardener. She started the garden in the rain of our first April here in 1985, and has been out in it ever since. Being in rural west Clare, and living the quiet way we do, some days meeting no other person, it is a garden seen by few. So, it is not grown and cared for with the goal of display. It is simply the natural extension of our living here, and as close as we get to having a relationship with the earth, not the one with a capital E, not the grand vast ungraspable wholeness of the thing that the steel turbines are said to be saving, but the portion of mud, clay and stone that live with us.

It is certainly true that at this stage, from all the weeding, feeding, seeding and planting in it, Chris knows each individual pocket of the garden. There must be some essential human goodness in this, in the idea that this parcel of ground has been in some small or large way enriched by our living upon it, and probably in the exact same measure as it has enriched us. I mean this not in a high-minded or self-satisfying way, but just the fact of tending the ground, feeding the worms I suppose, and that there is something absolutely right in this, and that the smallness of it, the most local of localities, the garden in Kiltumper, is enough world, enough earth, for a lifetime.

'I stand for what I stand on' is another quote of Wendell Berry's. And maybe because of the turbines coming, all the traffic and change that will happen just outside the hedges, I am aware of being turned more than ever to the world inside them, and allow myself to adopt the belief that for us *that* is the world that counts.

No sooner had I planted the fuchsia than Chris came to me with that gleam she gets that I know means she's seen something in the garden. It can be small or large. It can be a first budding, can be a plant thought lost that has come back, the first song of the cuckoo or the return of the swallows to the roof of the open cabin. Today it was a simple thing.

'I found two flat stones,' she said with a smile. She could have been her girl-self in the woods around where she grew up in Katonah, New York, hunting Indian arrowheads.

'Two flat stones?'

'Yes.'

'Where?' It was the least I could do to play along.

'Just past the polytunnel. They're buried a bit, but they look good.'

Good for what, I thought? I knew this translated as: *Let's go and dig them up.*

'Where do we need them for now again?'

'On the top of that wall.' She pointed to a dry stone wall we built that runs for ten feet along the eastern end of the 'join' that goes between the end of the house and the Glasshouse Garden.

I say 'built' while fully aware neither of us are stone-wall builders in any real sense. But over time, Chris, with her artist's eye, discovered a kind of serenity in the slow work of making a wall from a pile of stones. And, in the early years, stones seemed to be what most of the ground was made of.

Now, that wall, to my, and I would think, most eyes, did not look like it was crying out for two flat stones to finish it. It didn't look incomplete, but completion is a vexed question for gardeners. It is both a joy and a woe that in truth no garden since Eden is ever actually *done*. And I knew it would be letting laziness answer if I said the wall was fine as it was. So instead, I looked at the wall, and considered it as a form, something like a sonnet that needed a final couplet, and let that defeat my reluctance to go digging.

'Alright so.'

I went and got the shovel. Chris showed me the place where the stones lay. I could see where she had unearthed the flat surface of each. She had a *There, now* look.

'Perfect, right?'

And here's a thing. When I looked at them, when I began digging them out, I fell into thinking that she was right, they were.

Soon enough though, I went further, and came to the strange and perhaps inexplicable thought that these two flat stones somehow *belonged* to the wall.

This is a feature of all imaginative work. You're trying to find the words that belong to the story. It's a hard thing to explain to those on the outside of the creation, the sense that there is an inner order, a way the story or painting *wants to be*, and you are in some way trying to serve that. Sounds pompous when you say it out loud, and generally I don't. You spend two hours of a morning trying to find the right word or phrase. If you're me, in order to move on, you eventually allow yourself a near-miss, a word that's not quite right, but you'll come back to it later. That later might be mid-afternoon, or two afternoons later, when you're out in the garden, digging up a white fuchsia or a pair of flat stones, say, but your mind still has the knot of that word or phrase that isn't right, that doesn't belong. And eventually you find it. *That's right*, you think when you type it in. That belongs there.

And belonging is a key theme for anyone who, like us, has moved from where they were born to a new place. Thirty-four years ago, the west Clare we arrived into was not what it is now. At that time, a time of wind-up telephones, of hay-trams and neighbours 'giving a day' to each other when there was farm work to be done, we seemed to be the only 'strangers' in the parish. Chris was a Yank. I was a Dubliner. We were 'blow-ins' and although we had moved into the house where Chris's Breens were born, the idea of belonging here was not so easily accommodated. It needed to be made. So, it was not until Chris began making the first of the gardens, digging it day after day out of the miry brambles and nettles, both resurrecting and creating, that she came to feel some sense of being, well, *at home.* Home is where you dig.

All this to say, I may have been predisposed to the notion that the proper place for Paddy Cotter's white fuchsia was inside the hedge line of the garden in Kiltumper, and the proper place of the two stones Chris found was on top of the dry stone wall near the house.

In a further step of magical thinking, while digging, I allowed myself to imagine that aeons ago, these stones had become separated from the others, that when we gathered the ones with which we eventually built the wall we had somehow missed these two. By lifting and placing them on top of the others now, we would not be building anew but restoring something older than ourselves.

Which may well be nonsense. But it's the kind of nonsense I like.

*

These days I am aware of a need to come to a new accommodation with the garden, for each moment it shows me we are not as young as we once were. My way of working has always been fundamentally different to Chris's. Where she spends a kind of timeless time going from one job to another – that she may only discover at that moment: 'Even the anemones will need support in this wind' – I have tended to pick one large job and go at it full on until it is done. In this way, I have worked hard, effortful, in a kind of gardening combat, working towards getting-it-done, that moment when you have a cup of tea and sit on one of the seats against the house, look down the slope and think: *There. That's done.* This satisfaction, which I know has folded into it a bogus need for order, for an illusory control perhaps, may be less to do with gardening and more my own battle with purpose. But, having just passed sixty, I notice how the garden is teaching me to take it a bit handier, to find a rhythm in working that is not always bent on trying to get the thing done.

~~~

Gardening is the new yoga, I tell N. Slow down. He has yet to find his rhythm. It's like breathing... Not really, but there's a steadiness to it that he has to master if he's going to keep his post as Chief Groundsperson. I'm a triple fire Aries so it's not in me to slow down but N is a Gemini.

The UK's Royal College of Physicians published an article: Gardening for Health, *and included eighty-one citations of results of clinical trials and studies in 'green care', or 'therapy by exposure to plants and gardening'. All evidence supports the beneficial effects on mood and mental health when combining physical activity with exposure to nature and sunlight. In Kiltumper we've got the exposure to nature full on (you might say even too much of it because we can go a day or two with seeing only trees and shrubs and stone walls, but not a living soul except the ones in cars driving past to collect their children at the local national school); and physical activity, well that's part of living in a rural garden in what visitors like to call the middle of nowhere. Everything requires us to act upon it. But sunlight? No. According to the records of Met Éireann, the Irish meteorological service, we get sixty days a year without any sunlight. At all. Nada. April and May and June see the most sunlight – on average, between five and six hours. In January and December we get one and half hours of sunlight. That's sad. That's what's called a Vitamin D deficit. That's what's called SAD, Seasonal Affective Disorder.*

The RHS says the cure to loneliness is having a front garden. (Note: not a back garden or side garden but one where the neighbours pass by for the chat.) N can be seen once or twice a week chatting with M when he passes by on his daily walk to the top of Moran's Hill. N leans his arms upon the gate. It's like their personal outdoor man shed. They talk about their children, the weather, politics, when is broadband coming, and these days, what's happening with the turbines. M is a great neighbour, and many is the time he's been called up to help us fix our pump, or solve an electrical problem, or give instruction on the strimmer that keeps cutting out. He seems to know everything about fixing things, which is a stroke of good luck for us because Himself knows ... books.

There is something in this business of trying to get a thing done that is tied up with time and ageing and the knowledge of your own mortality. In it too is the realisation that although every garden is always saying *Now! Now!*, at the same time it is whispering *Slow. Slow.*

In this, I am lately reminded of Joeso, Chris's cousin, who was a quiet and gentle bachelor unhurried in all things. In our first April he came one morning to help us 'get in' the potatoes. He came carrying his fork. The handle of it was the first thing you noticed, the handle had a sheen to it and that all-over tan colour that evolves from the close relationship between man, tool and ground when that relationship is balanced. (This was a time when one measure of my own foreignness to the life I had leaped into was how often I bent the prong of a fork out of shape, a shameful thing it seemed when I brought it down to Tommy the blacksmith in the forge, because it showed lack of respect for both tool and ground.) Well, Joeso came in his suit jacket and trousers and strong boots with new yellow laces and looked at the ground that needed to be opened, and without taking off the jacket or laying a twine or otherwise marking the line he began with the fork, and *with apparent effortlessness* turned over the first sod, and the next, and the next. There was a mesmeric, metronomic quality to it. Along he went, a kind of Kiltumperian maestro, opening the way in a perfectly straight line. At no point did he take off the jacket, at no point did he say *Desperate ground*, or *Isn't there a world of stones?* He just did it. When he was done, he went back down the road to his tea with his sister-in-law, Mary. He had grown up here in this our house and he must have silently conversed with every particle of soil at some point.

And so, *Joeso be with me*, was my thought as I went digging out the two flat stones for the wall. Take it handy. The stones were large, and it was true for Chris that they were too handsome to be left buried. I unearthed them and brought them over to the wall. I didn't place them on it, of course. I lay them on the ground and Chris came over and looked at them and looked at the wall, and then looked at both some more.

'Try that one there maybe.'

I did. We looked at it.

'A bit over there. That side out.'

I did that.

'Now this one, here. Try that.'

I did. We stood back and looked. It was the late afternoon. Nothing was coming or going on the Kiltumper road. In one way I was aware of the smallness of our act in completing the wall when measured against the many older stone walls that would soon be bulldozed just outside the hedge line. But in another, I had a quiet and deep sense of victory, the wall looked, well, better, and Paddy's white fuchsia was safely in the ground in three places.

This is what matters this day, I thought. This is the way we are living.

We both stood and looked at the wall.

'That works,' Chris said.

*

A landscape that goes undescribed becomes unregarded. This idea in Robert Macfarlane's superb *Landmarks* leaps out at me this evening. This is the very thing that underlies all this describing. *'Regardez!'*

*

Today, in touch with my surroundings, literally. We are outside working for five hours. Mucky work really. But again, so much of gardening is about what you must do now knowing what will be to come. There's this loaded anticipation in it, in which it seems is the oldest covenant between man and earth. The faith that yes, a spring will come, and that, if cared for, the ground will once again offer up all that is needed, and more. So, we must prepare the beds; is it in Shakespeare that spring is compared to an honoured guest, or have I made that up? Either way, the image seems right as there's so much clearing out and cutting back and turning over to be done, a whole programme of making ready for something important. Chris itemises the various things that she has been asking me to move or divide all winter, and now, *Really now!* is the last chance we'll get, she says.

Right so.

It is work that I like doing, enjoying any job that has definition – which these days I am more aware is faulty thinking, whereas, as I've said, Chris might be doing a dozen different jobs at any given moment; I haven't that capacity, that whole-garden view. Which may be just as well, makes for complementary gardening.

I do have a different way of working in the garden than N does. Whether it's my ever-present sense of so much to do, where do I begin? Or, as my children have sometimes suggested, it's my tendency to go from one thing to another like a bumblebee. It's called multi-tasking in my book. One minute I'll be snipping spent flowers from the rose bushes (N doesn't always keep up with it, though it is 'his' rose bed) or I'll see a stone out of place along the edging wall, up towards what we now call 'The Wedding Field', where our daughter Deirdre and Nik got married, and I'll be on my knees realigning not only that stone but then all the others along the path. And then I have to weed around those stones. Dandelions and lesser celandine grow crazy here and need deep-down-into-the-soil weeding to get the full root. Or, now, over there in the pea-gravel the orange poppies (Papaver somniferum 'Orange Chiffon') have self-seeded in abundance and are begging to be transplanted to that place called elsewhere. A job I can do only while they're still just tiny seedlings.

What were you doing today? N will always ask. Pottering is always the answer. I was pottering…

Anyway, today, much lifting out and dividing. Mud and worms.

Several times in my garden notebooks, as if underlining for myself something I keep forgetting, I have written: *No garden is ever done. No garden can ever be finished, can ever be perfect.* And twice I have added: *And this is marvellous.* Think how terrible it would be if you stepped outside, looked around and concluded: nothing needs doing, the garden is perfect. It would be the end of the relationship, and by extension our living with and on the planet. So, no, a garden cannot be done. Now, while holding this idea, the trick is still to do the work for it each year. You do it, aspirationally, I think, knowing that there will be failures but still doing it anyway, sort of like living.

Several hours of digging, lifting, handling roots and earth today, all still in that dreaming of the garden to come. Throughout, there's plenty of what Wendell Berry calls 'the necessary enactment of humility'. By the end of it, at four o'clock, by aches my hands reveal all the intricacies of their bones. Trying to get off my gloves, I have a plethora of pains, so I don't know whether to save the knuckles or the wrists or the knots at the base of the thumbs. It feels not so much that I have been in touch with my surroundings as my surroundings have, in the firmest of grips, squeezed to crushing my hands' hundred bones.

'Can you help me get my gloves off?'

Chris tries to pull one off, it's not easy, and for a moment I can see the picture of us through the eyes of anyone who might be passing the gap in the hedge, this older couple in muddy clothes and wellingtons, wind blowing Chris's hair as she tugs a glove over the base of my thumb, and I groan a little and she laughs and I laugh, and there comes to me the fantastical thought that gardeners grow into the shape of their gardens. That our bones are become like the branches of the tree peony over under the sycamore, or the knotty twists of the cherry, that after all this time in this space we have grown in it the same as everything else. It has entered our skin, our bones, and then our spirit, until we are one. Maybe this is just the natural extension of the idea of sense of place. People can become a place perhaps. Perhaps, after all this time, as with the Vineys in theirs, this garden is us, not only spiritually but physically too.

The natural corollary of that is, when we tend it, we tend some part of ourselves.

Does that seem far-fetched?

When my gloves come off, the bones of my hands seem to want to stay curved, as if around the handles of an invisible barrow, and as I come inside to wash they stay like that, the barrow always just ahead of me.

*

I am not sure there is a word to capture exactly the sense of walking the road in the rain in Kiltumper. This afternoon, Chris and I decide the soft drizzle is as dry as this day is going to get. 'We must get a walk in,' she says, and I understand there's more than exercise in that imperative, and that among other things is the need not to feel trapped inside the bars of rain, escape her now chronic pain, and the various impasses in our writing.

We go for the walk most dry days, and a good few wet ones as it turns out. Today, the drizzle is almost nothing, almost air with its atoms showing. It is not falling, it is a stuff, a matter, visible and tactile, that hangs in the air and just above the green of the fields.

We get a good way along, past the avenue into what was known as Lena's, after the woman who had once lived alone in the house there. She had died by the time we arrived, but one story we heard in those first days was that Lena had the first piano in the parish, and I can remember going up to the ruin one day and climbing in past the brambles to discover it was true: the brown and dust-covered skeleton of the piano was sitting in the parlour of the falling-down house, its music no less lovely for being silent now. I know that's a tricky one to explain and may be a twist of imagination too far for some. But there's something moving to me in the way people leave their mark on the landscape. It may be true for all rural places,

47

I don't know. But one of the notable things in the human history of this place, is firstly, how common it is for a family name to be given to a crossroads – Shaughnessy's Cross, for example, just south of the village – and for that name to still be in use years after the last Shaughnessy has gone. At the eastern end of the Kiltumper road is Commodore's Cross, but the Commodore Conway was dead long before we got here, and he was not a commodore. Then, there are the fragments of story – like Lena's piano – that outlive them. This is the same as the story of Chris's great-grandfather Jack-of-the-Grove, riding his horse bareback around the parish, in the pockets of his long coat apples that he would set rolling on the table for children. It doesn't matter if Jack only did this once in his life, that is the fragment of story that landed in the mind, that connected his character and his time with this place, and, in time, both outlived and captured him. That fragment is all that is left. But when you stop and think about it, it opens up, the way a detail in a story can, and the manner and attitude of the great-great-grandfather are suddenly there.

Well, I love and am moved by this kind of local, I'm going to use the antique word, lore. The small prints of story people leave in the landscape, and which will endure as long as there are people living in it. I have thought of this lately too in relation to Chris and I. For a time, I expect, when dead, I will be remembered hereabouts by a shrug and the phrase, 'He wrote books', the content of which will be unread and soon forgotten, but *He wrote books* is good enough for me. But of Chris, it will be said, 'She had a lovely garden', some of which, no matter who lives here after, or if it all goes to wilderness again, some things will be living still and seeding and scattering and appearing who knows where inside the hedge line and beyond. *She had a lovely garden*, I thought, would be a worthwhile fragment in the story of place here.

And all of this is part of our walk today. When you've walked the same road for more than thirty years it no longer has just a physical dimension. You're walking in memory and history; you're walking in the moment when on this same road you first let go of the back of the pink bicycle when your daughter sailed away from your hand and in fierce focus pedalled free for the first time. You're walking on the day you walked behind her in her wedding dress when she was sitting in a pony-and-trap that was being led by a top-hatted great-coated Tommy Ryan through the drizzling rain up to the Blessed Well for her to be married. You're pushing your son in a stroller whose best days are behind it and whose wheels never dreamed of a road this rough; it makes his head bob, but he likes it, and it's the same road he encourages you to jog along with him a quarter of a century later when it's your head that's bobbing and your bones jarring from the same roughness. You walk along it holding your wife's hand when you hear he has passed the New York Bar exam and will become a NY attorney-at-law. It's the road you both walked with Huckleberry your dog for his fourteen years. It's the road you and Chris walked sometimes weeping, and sometimes without words, when she was trying to come through the chemotherapy.

There are an uncountable number of memories here, we have populated the place with them. The thing, I suppose, is that the road has become *personal*. It's a public road, it's not ours in any real way. But a significant portion of our life has been here, on its two-and-a-half-mile length we have literally walked thousands of miles. And that matters.

Now, Lena's is where the wind turbines are going, and without saying so I've noticed in recent times that as we are passing that part of the walk we look south, away from the wounded place of what is so far only a site. In future times, this will have to become habitual, keep your head turned to the south and not to the hill to the north where the blades would shred our peace. If we can't manage that, if we can't make peace with it, we won't be continuing to walk this road. Already Chris is walking it less, and it's a sad fact of life but this is not the stage of life for her to stop walking.

The road dips and we pass the stretch of forestry and then the two still-standing walls of the once-upon-a-time tiny cigarette-and-sweet shop by the river. Then up the rise where the western wind is always vanquished by the geography and a parcel of dead air and stillness inhabits like a ghost site. We go on past the empty house shell with the best view in the parish. It looks out down the valley towards the village nearly five kilometres away. You can see the church spire, just. Beyond it you can see the mountains of Kerry. To the west you can see the rise of land that falls away into the ocean at the cliffs of Kilkee.

The walk always has the natural destination of the view of the sea. By silent but mutual accord, we go until we come to the highest point of the road, from where we can see what is sometimes the dull grey and sometimes the gleaming silver of the Atlantic. We see the sea, take a breath, and turn for home.

We are not far from the crest when the drizzle turns to a steady rain. It's a rain that announces itself, that's not to be mistaken for anything else, goes from soft to hard quickly, so that soon enough it is running into the pockets of your coat and painting dark the thighs of your trousers. Your shoes hold out a little longer, then, sinking boats, surrender.

And now Chris and I are walking along in a downpour. There's nowhere to go for shelter, you're already wetter than makes no difference, so there's nothing to do but go with it. We're a little over a mile and a half from home, so what of it? No cars come or go. We carry on to the crest and take a look at the no-view of the not-seen sea, the rain coming down so straight and saturating that when we look at each other we can only laugh. Which, to two sixty-something-year-olds standing in the middle of a country road in the west Clare afternoon, is one of the gifts of the rain.

We turn and walk back. And what strikes me is the quiet, that is not quiet exactly, because there is the noise of the rain, falling and landing, a noise small and shushing at the same time. It makes no sound at all as it falls on the pine trees of the forestry, and then they make the dripping that seems unreally distant because you see but don't hear it in the acid carpet of needles. In the fields the grass swallows the rain without comment. The falling from the sky is so steady and enveloping now that it becomes a kind of quiet. It has its own substance, sound and texture, but this afternoon as we walk along in it on the Kiltumper road, saying nothing, it feels we are moved into some other, dare I say – I do, and will – mystic state, where all the times we have walked on this road are present, and all the earlier versions of ourselves over the past three decades and more are too, and there's some profound tranquillity in us at that moment, wet and all as we are, and will be for some time.

⤙

I am reminded of Flann O'Brien's The Third Policeman, *which I studied in UCD when I got my MA in Anglo-Irish Literature in the last century. O'Brien has an hilarious section where the*

*sergeant is telling the narrator about Michael Gilhaney who is –
because of the atomic theory being at work in the parish and
half the people suffering from it – nearly half bicycle. Because,
as O'Brien writes, 'Everything is composed of small particles of
itself, and they are flying around in concentric circles and arcs
and segments and innumerable other geometrical figures too
numerous to mention collectively, never standing still or resting
but spinning away and darting hither and thither and back
again, all the time on the go.' Resulting, you see, in the atoms
between the bicycle and Gilhaney commingling and becoming
each other.*

*Just so, we are part of the road. We and our children have left
our atoms on it and it has become us. Joeso was part of the garden
and the garden part of him, and Mary Breen, and Jack-of-the-
Grove, and on and on.*

<hr />

Today the birds are remarkably alive with song. There is a
telling in it, full-throated, urgent. Lift your head and pause
and it's like an actual sound-cloud all around you. There are
still no leaves on the trees, but there is light coming. There is
light coming.

Primula Denticulata

3

March

Time and the garden.

A peony, I found out today, can live for a hundred years. That bears thinking about, and I was thinking about it all afternoon when I was outside spreading some bark mulch beneath the roses and alongside two fairly gnarly-looking peonies that Chris has told me are growing in the wrong place. This idea of 'the wrong place' I know I will keep coming back to. But, looking at the peony, time was what I was thinking of. Seemed kind of extraordinary to think of a hundred years of trying, each year, to bloom.

Another plant in the wrong place? I'm afraid so. The two tree peonies look like gnarled sticks, growing crookedly to the west, trying to escape from under the shade of the sycamore tree above them. They need a space of their own. Some plants need their space, they need to be admired to grow properly. (Some of us need more attention than others.) Some plants don't mind where they are growing. Like the yellow loosestrife (Lysimachia punctata) which really is a pretty plant but it, like the Japanese anemone,

likes to take over. Another one for the ditch across the road. If we move those two, I tell N, then we can 'shift' the tree peonies into their spot in the front bed to the east of the conservatory near the delphiniums. Except, they don't take well to being transplanted. So, do I try, or not?

In the lower bed, there is a ginkgo biloba. We planted it more than thirty years ago, and after all that time it is only twelve feet tall and thin as three fingers. Its branches are less than a foot long. In fact, the tree grows by inches only, and only upwards towards heaven. I think I remember it is a tree mentioned in the Bible and one of the slowest-growing trees on earth. Its fierce focus feels monastic and makes me think of single-minded devotion. It is not far from the flowsy silver-leafed pyrus and seems to reprimand it for its theatrics. Pyrus has outgrown the ginkgo by eight feet and is twenty times broader with the stretch of its stems. Ginkgo doesn't care, doesn't care how long it takes, time is nothing to a ginkgo.

So time is something a garden keeps redefining in plant terms, not human ones. They say it takes thirty years to make a garden. By that I take it to mean it takes thirty years for you to realise the mistakes you've made starting out, thirty years for the land itself to teach you what goes where, what can not only endure but thrive in this spot or that one. Thinking of this today, looking at the trees in the front garden, all specimens chosen with some thought by Chris, who hates the idea of getting something 'wrong', chosen so that the trees 'spoke' to each other – the ginkgo's steady aspiring, the pyrus's leaf carnival, and over beneath the wind-shelter of the stone cabins, the Mount Fuji flowering cherry, the star of the garden, the one that for three weeks in May draws the

Oohs and Aahs from those who pass in through the stone archway and enter literally the 'outdoor room' of the garden.

I'll pause here to say something of that doorway. Well, as I've said, the line of stone cabins was once the living quarters of Chris's great-great-great-great-grandparents and their various livestock. The cabins run north-to-south and face west. The largest cabin, about fifteen feet by eighteen, has the exact dimension and layout of our main room in the house, and was built in the eighteenth century. The cabin has a stoned-in chimney and a loft ledge across from it where once no doubt a number of Breen children slept smoky sleeps. When we first arrived here the advice we got was to knock down the old cabins completely – 'It would allow you drive right to your front door. You'd get a great sweep. You could drive in, right around, and out the back!' Advice we nodded politely to and didn't take. People can be quick to knock things down. Well, eventually we took the stones out of where they had been carefully built into the doorway and opened a passageway into the garden, which has become the way we come and go ever since. This has several advantages, the aesthetic one of coming through a dark space of stone into a green one, the theatrical one of entrance, the historical one of coming through the ancient into the *now! now!* that all gardens shout. But also the gardening one, because of the cabins that were not knocked there is a line of shelter from the Atlantic wind, and with that one wall of a walled garden, in this case ten feet of stone with fifteen of roof, we can grow things that would not survive just the other side of the cabins. (I'm not kidding, just the other side of the cabins that wind of salt and hard rain ate the haybarn. First it sampled the edges of each panel, rusting the joins, biting loose slabs off in the winter and making the whole haybarn sing a tormented ache-song. A song that grew worse when I went out with a blue rope and tried to tie down the loose panels, lending the haybarn the look of a forlorn shipwreck, full of criss-crossed rigging, and giving the big winds something new to play. You can't defeat the wind here. The tightest knots I could manage were undone in days, in the morning the haybarn waving blue ropes gaily at the kitchen window. At last we surrendered and took the barn down.)

Well, because of the protection from the cabins, we were able to grow the three trees, but especially the magnificent, almost horizontally flowering Japanese cherry. Each year its blossoms held for a few weeks only, but their sheer beauty always won the argument as to whether it was worth waiting a whole year for just that brief display. The tree spoke to the very opposite of the ginkgo, to the single moment of flourish. For me, it spoke of rapture, of inspiration, which may take a long time to come. Each May every inch of the cherry would become heavy with blossom. You'd stand underneath and be in a communion-time of whiteness, an otherworld. You'd look carefully at the wood of the branches. It was hard and rough in places with some fringes of a greenish grey lichen, and nothing about it suggested that from it would burst these flowers. It was just pure wonder.

A week or so after its peak, the tree would start releasing the blossoms and we'd have a white snow flowing across the garden. This is the thing that is the important part of their role in Japanese culture. It is not the flowering, but the falling of the flowers, for it speaks to time and to all of our passing. 'Things reveal themselves passing away' is a saying from somewhere in Yeats, I seem to remember, and seem to remember too that late in his life he had an affinity for Japanese culture. So, there is that too. And of course, as is inevitable when you reach a certain age, that awareness of passing is never far. In any case, the cherry's is the most beautiful dying, and there's something profoundly sad and hopeful in that at the same time.

All of which to give some indication of how sad we both felt when we discovered in the spring that, after thirty years of growing, the cherry had died. As always, Chris noticed it first. I didn't believe her. It's a characteristic of mine to resist reality. 'It's just taking its time.'

'It's dead, Niall.'

'Couldn't be. It'll come. Just wait.'

'All right. But I know.'

And of course, she did know. Even a fiction writer can't defeat reality all the time. The spring came on but no leaves appeared on the cherry. Its time was over. But I can't quite bring myself to cut it down. That tree has been a companion to all our years here, there are many photographs of the children standing beneath its scented canopy, many memories of watching the blossoms snow into the grass. I know it will have to go, someday, but – absurdly I know – I feel we should give it this summer and consider removing it in the autumn. There's something about taking down those thirty odd years and planting anew a tree that will be in this garden for the time after Chris and I are gone that chills me a little, and I don't quite have the courage just yet.

Give it time, the ginkgo says.

*

Today, news of a storm that is headed in across the Cliffs of Moher. The radio warns us we are Status Red. Winds will be over 150 km an hour. Stay indoors.

We are outside immediately after breakfast. The garden is strangely supremely calm. There is not a breath. There are no birds singing. The absence of their song is startling, and you miss them more than you appreciated them yesterday. *They know*, you think. Though the storm is still some miles off the west coast of Ireland, and the birds of Kiltumper are not coastal or seabirds, still, they know. I look up into the branches of the just newly leafed trees and see no birds sitting there. *Where are they gone?* The feathered plumes of the old pine trees across the road are not moving at all, and I think, maybe the forecasters have it wrong, that the storm churning now out in the Atlantic will turn north and miss us entirely.

I think this but don't say it to Chris, because already she is gathering bamboos and string, and it will seem at best lazy or at worst uncaring if I say we don't need to worry about the plants, many of which are now a tender eight inches above the ground. The first hours of the morning are taken up with staking and supporting the spring garden. Chris ties each plant, like sailors to masts. She misses none. I hand her the sticks and she sinks them carefully, avoiding the roots. We had a mild last two weeks of February, so growth seems ahead of itself. Green string goes around each. There is phenomenal care in this, and it's not a leap to think of it as a kind of practical love. It is moving because it is her instinct, she couldn't do otherwise, leave any plant to chance.

And it is just as well. By the time we have finished, the tops of the pines are swaying. I go inside and ready candles and fill pots with water. The storm is now scheduled to make landfall in Clare at four o'clock, the radio says.

At one minute past four we lose the electricity. Big-shouldered and broad-chested, the wind throws itself around in the garden. From the window, Chris sees that the support for the baptisia has already come loose and she goes outside in the hurly-burly quickly to reset it.

The storm proper arrives by six, after darkness has fallen. By eight it feels as though the world is blowing away outside. Branches, old, and leaves, new, come flying at the house. The dark is both a mercy and a fear, as a blind ocean of noise whistles and roars and sticks start to clack one after the other against the windows. The immensity of it is what is startling. And the force. It is elemental and raw and immediately makes you realise the smallness and vulnerability of us all.

'How will anything survive in this?' Chris asks, and I know some part of her is out there in the garden.

'They will.' I don't believe myself. 'Most will, anyway.'

The house sounds like we are on the sea and in the sea. We have had many storms, but by a clemency inbuilt in life you forget most of them after they have gone, and when the next one comes it seems to stir up a sleeping anxiety. The house has come through 200 years of Atlantic storms, its walls are stones piled three feet thick. It survived the Night of the Big Wind in the nineteenth century, and the more recent ones that each year seem more powerful, whether from climate change or because as you yourself get older all defences seem frailer. There is no fear of this house even though it seems the windows will give in and the roof slates blow off. This is what I am thinking, and of the practicalities of water, heat and light, but I know that Chris's main thought is the garden, and it is not stretching things to say in this she is like a mother whose children are out there in a dark hazardous elsewhere.

In writing this, I am aware that to anyone who is not a gardener this may seem ridiculous.

To anyone who is, it will seem natural.

The storm rages and bangs and howls on through the night. She is called Freya and is enormous. We sleep in fits and starts, and each time we wake it is still blowing.

In the morning, silence.

Freya has gone. What we see when we come downstairs are the windows to the west completely plastered with small leaves, like tiny green palms pressed against the glass. There are hundreds of them, new leaves shorn off. The gravel outside is covered with small branches and carpeted in green leaves. Chris and I go out for a survey. I am looking at slates, she is looking at the blow the pyrus has taken in its midriff, a definite hole in its silver belly, its innards littered on the grass. But for the most part, the canes have helped greatly. The soft new growth has held on to the twine supporting it. The spring has not been destroyed. I gather barrows of fallen branches. *C like a nurse in a ward*, the image I think to put in my notebook.

In the trees, the birds are back. *Where exactly were they?* Well, wherever, I don't think I have ever been so glad to see and hear them.

There is no defence that N can offer. Somehow, he has accumulated five pairs of work gloves. Washing them this evening in a basin, I needed to change the water three times before it stopped looking like a muddy river.

When I hang them on the clothesline between the sycamore and the birch tree, they wave at me, ten hands of air that will be back in the earth before long.

The purple-sprouting broccoli are sprouting like nobody's business. There is an absolute abundance of them. They are the first we have ever grown and in March fresh green vegetables are twice as valuable as in summertime. Their taste. What can I say of their taste? It is one of the truths of gardening that all home-grown vegetables taste better, how could they not? And maybe just how much better is quantified by the amount of labour they have taken and the care they have needed. Perhaps there is an unwritten magical formula whereby sweat is converted to flavour. The goodness of the taste is inextricable from the knowledge of goodness itself. Whatever we might mean by that term, it seems indisputable that there is something 'good' in growing, and of course, that informs your taste. So maybe I can't fairly judge the broccoli. Well, that doesn't matter a whit, as we sit down to eat a feast of them this evening.

'Beautiful broccoli,' Chris says. 'The water was purple!'

Remembering the Christmas morning when I had given each of them their sup of Harp, I add, 'And not a whiff of the beer off them.'

*

What is it that makes your heart lift in a new pair of wellingtons?

Does that sound absurd? For three months out the end of last year I was wearing wellingtons – expensive gifted wellingtons – that had separated seams in both boots, so that if I stepped into moist ground, of which we have plenty, a bleed of water came over the seam and soaked your sock. Now, as with all things, to those with my nature at least, the first step was not to throw them out and go get a new pair. Once I became aware of the leak, I negotiated a way around it; with caution and thought I worked around the infirmity of the boots; after all this time they had become very comfortable, in fact *they were me*, and I didn't want to surrender them. So, if I had to, I could heel-walk across the wet places, and if it looked odd, what of it, it saved your socks, and who was looking? There are some privileges in remoteness. One of which is you can wear anything, and act anyhow. You might as well enjoy it. So, yes, heel-walking. And only occasionally, pushing a loaded barrow say and seeking the shortest route, to hell with it, you went as fast as you could through the muck and puddle, thinking speed might save you, (you could get across before the ground *knew* you had punctured boots,) and shortly felt the cold sensation of walking on water.

You must get new boots, you say to yourself. But you don't. You'll manage another day, and another, and in my case at least, get used to taking off soaked socks as you come in to lunch and discover your feet cold as a corpse's.

The day to go and buy new wellingtons, I don't know what day that is. It never seems to come. I have worn wellingtons with bicycle-tyre puncture patches stuck on, wellingtons with duct tape, wellingtons with blue plastic shopping bags inside them.

This to measure something of my delight when Chris gave me a ribboned, wrapped box at Christmas, in it a new pair of wellingtons. I am aware it might seem a stretch for anyone to imagine my joy in getting them, but imagine on, a joy is a broad company. And soon enough, there I was, walking in them across the Grove, realising that my body's instincts had become trained to avoiding the smallest gleam of wet ground, that now I could choose to cross. And in doing so, experience an instant spirit-lift that was innocent and boy-like. I couldn't help smiling. As though waterproof boots had only now been invented, and here, in the garden in Kiltumper, is a man in his sixties marvelling at that, and well, yes, puddle-dancing.

*

Home truths: I know no other way to write other than the personal way. And it is always about love.

Chris knows no other way to garden. It is personal to her. That is to say, it is about love. It is always about love.

*

The garden is its own social network. Utterly interconnected. Chris knows this instinctively, me intellectually. So it is always a surprise of small enlightenment when she points out how something has benefitted from its neighbour. Or been overshadowed, or strangled by it.

I, of course, am essentially a solitary man, and prefer the habit I grew into – perhaps not exactly by choice – I always imagined many friends – by some essence of my temperament, nature, I suppose. And something of this you learn in the garden. Gardening is by and large solitary and silent. And though, over there by the cabin bed Chris is working, she is solitary and silent too. And the solace of that, of her being there, is for me immense today, for, more keenly perhaps now than before, I know it will not always be so. I put down my trowel and go across and hug her with my wrists, her and my muddy gloves sticking out at our sides like unformed wings.

'What was that for?'

'For being my social network.'

Another appointment, another visit to the consultant in Galway. These appointments I can go to alone. N has accompanied me on so many visits already. I hate to drag him along. So I bring 'my bloods', tell the bone specialist how I'm feeling. I tend to rattle on about the beliefs I have that the daily injections

are affecting me, negatively. (Teriparatide. It's a high-tech medicine in the form of a synthetic parathyroid hormone to promote bone formation in severe osteoporosis.) I don't think she buys any of it. This is the protocol for someone at 'high risk for spontaneous fracture,' she says. 'It's only for two years and you're midway there.' She asks, 'Are you doing your exercises?' 'Yes; sort of. I mean I work in the garden for hours a day,' I say. 'You should see me in the glasshouse balancing on one foot (as there isn't room for two feet on the little ledge of block), making a structure for the tomatoes out of bamboos going horizontal and vertical and tying them together with thick cord.' I go to Kevin's excellent t'ai chi classes in the village, not as often as I should, I know. Some of the exercises seem too easy for someone who has built a stone wall.

'I hope you're not building stone walls now!'
Well…

It's highly disturbing. I mean after coming this far through the cancer and all, and still the continual surveillance as I call it, I have to deal with severe osteoporosis. If I'd understood sooner that although I have a family history of it, the fact of there being no Vitamin D available here in Ireland from the sun during the months from October to March, I may have paid more attention to taking supplements when I was in my thirties and forties. I may have insisted on a sun holiday in the winter because it does store in your body.

Chemotherapy depletes calcium so I'm snookered there too. I'm hoping for some luck when the injections cease; the bone density of my porous bones should increase. Otherwise … well, it doesn't bear thinking about. I'm not going to think about it. I'm going to go outside.

Here is a moment in the history of this garden.

It happened that Chris and I were invited by the University of Notre Dame to a weekend retreat at Kylemore Abbey in Connemara. We went, partly out of the novelty of invitation, and partly because of the irresistible beauty of Kylemore and Connemara. Well, when there, we met Father Tim Scully, an enthusiastic Irish–American with smiling eyes and the kind of shining some people call the holy spirit. Tim had an irrepressible energy about him, had read our early books about moving to Clare, and so when, a few weeks later, he emailed us to say he wanted to stop by Kiltumper to see the garden for himself there was no other possible response than to ask what day.

He came on a day of sunshine. He wore a short-sleeved shirt and a wide-brimmed straw hat, so he seemed a visitor from a distant land, and when he came in through the open cabin he stopped and exclaimed. He opened his arms towards the garden and just held them like that for a moment, looking across the front beds and down to the lower ones, palms open, before he turned a pink smiling face to me and said, 'I'm in Kiltumper.'

Well, we felt kind of honoured, I suppose, as we always do when we meet readers who have been here in mind long before they arrive in body. There's a strange kind of connection between the reader and the place written of, and it always stirs in me both the fear that the place will not live up to the imagined, not imaginary, version, and a sense of a completed circle between writers and reader. So, with Tim, and his assistant Matthew, there was certainly that. We had Chris's sister Deirdre visiting, and had planned a simple lunch with soup and a salad drawn from the garden, but just before we were heading in to sit down Tim said, 'I'd like to say Mass in the garden, would that be OK?'

Now, Chris and I were both raised Catholic, and for various reasons both simple and complicated, shallow and profound, have fallen out of the practice of Mass. It is one of the oddities of my nature that I feel more alone when in a congregation, and Mass in the parish church became a deeply lonely experience, so I stopped going to church when there were other people there. I go on my own, into churches I happen upon. But the fact is all of my fiction is about faith in one way or another, and some powerful underground river of it flows through me still. So, when Tim asked, I looked to Chris and we both nodded more than said OK.

I'm not sure he was waiting for our answer. Tim thought this was going to be marvellous altogether. With a lively step, he went to the car and came back with a small case and his Bible. He set the case on the metal table outside at the top of the garden, took out a travelling chalice, a small bottle of wine, and asked Chris for a little water. While he was waiting for her to come back, he beamed at the garden. 'This is my basilica,' he said. His delight was infectious. There was an innocence about it, untrammelled by age or experience. Father Tim must have offered Mass tens of thousands of times, I thought, but he still had this spirit-gleam for it. I envied him, and felt I didn't want to let him down. He set the chalice and the Bible just so, and when Chris brought a little silver jug of water – I think a Christening gift to one of the children years ago and our only one up to the job – his smile appreciated the thought. He took off his straw sunhat and put on the stole. There were bees buzzing and some songbirds in the weeping pear tree, but no other sounds. Nothing was coming or going, and the idea of the Mass in the garden immediately felt like some secret sharing, and amidst all that was growing and blooming, somehow right.

Tim showed us the readings and asked us each to do one. Then he blessed himself and began. It was one o'clock in the day, sunlight, and a small breeze travelling around the garden.

Mass outside in nature, in a congregation of five, has an elemental air. It was bare, and rudimentary, and *present* is the word I think I would use. Later, I would realise it was quite likely the first Mass ever said in that garden, but that if Mary Breen was right when she told us some of the lore and names of the fields, and that a corner of our meadow behind the house was in fact called *Pairc na moine* – 'You know, Niall, the field of the monks' – maybe open-air Mass had been said nearby in ancient times. And not too far away, down the road past O'Shea's and up on what is called the Scragg there was a Mass rock from penal times, and certainly people had stood and knelt and bowed there to a priest with chalice and Bible. Later, I would link this moment to those others and find that sense of rightness I felt had roots in the place. But when the Mass was actually happening, when it was just us in the garden, there was no time for finding resonances or thinking of history, we were in a kind of absolute concentration, charged with intention, aware that the Mass only happened if we gave ourselves to it. We were not only part of it, *we were it*. There was this feeling of bond, of the five of us making the offering there in the garden, saying the words, passing the chalice. When Tim raised it and said 'Do this in memory of me' I felt the words pass into me, the hairs on the back of my neck tingle, and I realised this was the most remarkable Mass of my life.

Tim did not offer a sermon. He just said a few words about his delight at being able to say Mass in a garden that showed such love. Later, I would think of the poetry of Hopkins, the world's greatest conjoiner of nature and God, but then I only thought of the word *celebrant*, and the celebration in it, and that what Father Tim was expressing here was joy. In saying the Mass he was joyful.

And here's the thing.

In the months since that day, I revisit that moment more than many others. Not necessarily intentionally. It'll come to me. I'll be on my knees at a bed, or barrowing Chris's latest mound of cuttings, and something of that Mass will be in the air and I'll think of it, how, though I did not go to Mass, the Mass came to me, and, because I am the way I am, my mind will make the leap and believe that what blessing or goodness was said or summoned that afternoon is here with us still, and we are more attended than we know.

Marsh Bouquet

4

April

The light. The light!

Every year the same surprise, the same lift in your heart as the days lengthen, and you look at the clock and back out at the garden and the sky, and it is as if your spirit comes out of a cave it did not realise it was in, because *Look! The light!*

It is always remarkable to me. Even after sixty years – or maybe especially after sixty years – the moment of spring sets everything in me tremoring. That sounds too theatrical for what is an inner stirring, but the idea of it is true. Having become grimly accustomed to the dark, I marvel at the April light and always feel a sense of having come through. The winter in the west is not perhaps as harsh as some others, but it is dark and wet and windy, the daylight shutting off in the early afternoon, the shut-in, isolate, and isolating sense of the landscape, and the older I get the more I am aware of the longing in me to make it to spring. This is maybe accentuated by the way we live here. When I was younger, busier in the normal way with the children or with teaching, I don't think I was as struck by it and was probably a few weeks into April before I raised my head and realised yes, winter is over. But now, just the two of us, all day in and about the garden, we are become if not exactly connoisseurs then certainly amateur authorities on the weather and the light we are living in. I say this not boastfully, but in acknowledgement of a natural outcome, how we have grown into this place and into a way of life. It couldn't have been other. Even for a concrete Dubliner like me. So, the winter is fairly *felt*, known at first hand as it were, and the first tentative moments of spring have that quickened and quickening sense of a pure renewal that you only now realise you had lost faith in.

For Chris, who has a technician's awareness of every change in the garden, the spring is in the earth not just in the air. Her eyes are aimed down, on the hunt for the growth. She knows the plants better than I know anything except maybe some books, knows which of the perennials will break ground first, which will follow, and what should be already showing. Because, like many mothers perhaps, she is a worrier, she worries that the first days of April have not yet revealed the shoots of the blue delphinium – or have the slugs already got it? – and she kneels in the bed and pries gently, only to come away, chastened but awed, like the doubting Thomas, having seen, *and felt*, the proof that yes, there is a rising.

This action of renewal then is everywhere in the garden, and so swift that it bypasses your ability to register it – or mine at least – and so always spring overtakes me. I cannot notice all of it, and one morning there is something already in leaf that seems to have unfurled overnight. It's rare that I should spot something before Chris does. And I don't mind that one bit. It's not a competition, April is fleet and various and the garden has many plants. I sit down outside, and in my notebook write: *Try and capture something of your own rapture in April. Make it earthy not airy.*

But I give up after a few notes, exactly because for me *it is airy*.

⌒

It is our thirty-fourth spring in Kiltumper, thirty-four years since that April Fool's Day when we first arrived from America with four suitcases and fistfuls of innocence and ten boxes of books. It was raining of course, and Mary, our ever-hardy, ever-optimistic neighbour, had promised sunshine coming with the cuckoo. Thirty-four years ago we walked into the overgrown jungle of the Kiltumper garden. The cottage had been vacant for five years and the garden, which previously had been devoted entirely to potatoes and cabbages and onions, was largely neglected. Our first fork and shovel have long since become artefacts, broken handles now victims of the hard stony ground in which we garden. Our daughter, who was an art student in Dublin, used to photograph their iron brown and rusted heads as they lay against the old wooden tram lift, up against the cabin wall, like objets d'art. *Inspiration for her fashion degree.*

Back then we had no idea of the rhythms of country life, and everything was for the first time. Although so many things have changed in Ireland, and so profoundly, it is in spring we catch glimpses of those rhythms still because nothing stops the growth of the garden. It's as if the soil says, time's up, up you go green shoots, take your chances, mind the slugs, strengthen yourself to bear the western wind, only open your buds when the rain has eased.

And speaking of slugs ... BBC Wildlife's website says that slugs are very important. Fodder for mammals, birds, slow-worms, earthworms and insects. Insects? Is that like David and Goliath? (Maybe there is an insect I don't know about that can tackle a slug.) Upset the balance by getting rid of them and you can do a lot of harm. I'll take that advice with a grain of salt, or two or three...

Yes, that's what happens on my slug patrols. I go out at night in my wellies, with a flashlight and gloves and pick them up and put them in a bucket of salted water. Slugs love Kiltumper. By day my slug patrol involves lifting the many terracotta pots under which they're usually hiding. I tip them into a saucer and sprinkle salt on them and then leave them out for the birds. If it's sunny, I'm serving roasted slug. If rainy, the menu changes to slug soup, which usually goes to waste.

Of the thirty slug species, you've got your great grey or leopard, which grows to 20 cm, and large black. There's a Budapest slug, and a yellow one, a garden slug – that seems obvious enough – a common grey field slug and a shelled slug. Two deterrents I will experiment with will be to make a gutter trench around a bed. Snails and slugs can't swim, apparently, and drown under water. And, like so many humans, slugs hate cucumbers.

Next up on the bird menu: slug on a slice of cucumber?

A few facts I didn't know, but which don't make me like them any better. They have green blood. Can live for six years. And worse – their eggs can lie dormant for many, many years.

One more thing, they can have 27,000 teeth.

I think my delphiniums have met every one of them.

⌒⥤

In this airiness is something of my own nature that I go back and forth on. My favourite cathedral is the long Atlantic beach at Doughmore, sixteen kilometres away. It is an absolute place of air and sky. For Chris, her cathedral would be a dense woodland like the woods of her childhood in Katonah, a green place, enclosed, sheltering, safe. Or equally, Montauk and Ditch Plains at the tip of Long Island before it became an 'in' place. As will be clear, I am the less earthy, more airy of the two of us; the actual air in the garden is a key component for me, and maybe more important than the earth itself. I know this is a kind of gardener's sacrilege and doesn't feature in any gardening book, and I have not come across any gardener speaking of the pleasure of the air in a place where things are greenly alive. And more than anything this may simply signal me not as a true gardener, but what I love is the moment when I unbend from weeding or clearing and feel the air cool the sweat I didn't know was there. The garden air then is actually delicious.

I know, *I know.*

I have no science to prove it, and reading these words on a page you cannot feel it, but believe me, it just is. Delicious. And in April, well. *April air and light, these are the ingredients of rapture* is what I write in my notebook, and don't care one whit if I seem doolally.

*

I should know more birds.

I will try.

But I garden with my glasses off, and too late see the blur in flight.

⚬

K, our neighbour up the hill, is moving cows along the road into the field across from the house, moving them from their winter shelter of sheds onto the grass for the first time. About thirty of them. All Friesians. The clatter of hooves on the road is steady and soft, an affirming sound of spring returning. K shuts our gate, an ancient iron and partly rusted, partly broken contraption, so the cows won't venture into our yard. It makes a scraping sound closing as he drags it across the pebbles. We seem to be perennially apologising to him that we haven't replaced it yet. The Friesians trot single-file through his gate down into the grazing meadow. The grass is ready for them. And they for it.

We saw his father do this when K was a small child standing alongside him, waving a short length of black piping. And one day, K's three girls will be herding them too.

A new word: *edaphic.* It means by, on, or of the soil. The unyielding 'f' and 'ick' sounds, the way it sticks too, are all present as I work in the cold damp ground digging out – '*Did you get the roots?*' – the last of the dandelions. *Edaphus*, it comes from the Greek for floor, and here on the floor of our small world we are silent and working in an elemental not to say essential way, floor-creatures thinking of what will arise.

This evening I write down the question that came to me while cold-fingered on the ground this afternoon: 'What does this piece of ground we call a garden *require* of us?'

That we are here *for it*, and not vice versa, is the turn my thinking has taken.

*

A garden has no existence without a gardener.

Some obvious things never occur to me. But today, watching Chris move about the garden, from one job to the next, all of which are occurring to her only as she sees them, what I think of is the way an artist chooses now this colour, now this stroke, without premeditation, in the *now* of making a painting. And watching Chris's version of this in the garden, I am aware of a kind of natural magic, a responsiveness, a reading of place, that I don't have. But which is marvellous.

An artist-of-the-garden, whereas I am more a colourer-between-the-lines.

She is purely intuitive. And because I don't have that intuition, or don't trust myself to have, there is a feeling of wonder about it.

This garden then is first and foremost an expression of its gardener, Chris. (I don't know if a garden can be the expression of two gardeners. I am more groundsman than gardener anyway, and happy with that.) It is her mark on this place, but it would not be the same garden anywhere else. In this, I am aware I am speaking against the idea of such things as the show gardens in Chelsea Flower Show, say, which have been transplanted into the showgrounds and so are I suppose artificially natural, if there is such a thing. The garden at Kiltumper could not exist anywhere else. If, as we have sometimes thought in the past year, we may have to leave here when the turbines come in on us, if we can hear them turning in the garden, in the house when the window is open, we know we cannot bring the garden with us, nor could we replicate it anywhere else, which makes the thought of that loss all the more potent. Because in some real and essential way, the garden is our life here. Now, we know that if someone else had been living here for the past thirty-four years the garden would not look like it does today. I don't mean to say not as beautiful. It is not a vanity, just an awareness of the interwoven nature of garden and gardener. Another gardener would have chosen different trees, shrubs, plants, most likely laid the beds in different configurations – who is to say? – the point is it would have been their expression as they responded to the space.

So, today, in our mostly silent, steady, coming and going between the beds, it strikes me that what we are engaged in is the making of ourselves, as creatures of the earth, with a small 'e', and what might seem paradoxical at first, but is in fact stunningly beautiful in the simplicity of its conception, in that actual earth-engagement there is, for want of a better word, soul-nourishment.

*N has been inspired to get out the mower and attack the somewhat wild-looking Grove. We like to keep it wild, but make some grass paths through it, through the daisies (*Bellis perennis *– no children left to make a daisy chain) and dandelions and buttercups and tall grass. The motor whirrs and zings in the quiet like a bell inviting summer. When the paths are mown they too are invitations, and I accept them happily.*

Bluebottle buzz has begun in the house. A bee knocks against the windowpane. More sounds. Birds continue to sing around the feeder. We await the sounds of tractors spreading fertiliser. Cows across the way are small smoky puffs of clouds, their calls, low and gentle, echo against the hills in the distance.

Ditches murmur with water. A new-born calf moans for its mother. A lonesome donkey somewhere up on the edges of the forestry brays. Wood pigeons – which always remind me of my bedsit in Sandymount when I was studying for the MA and my 'courtship' with N began – coo from budding sycamore trees. N wonders if the sparrowhawks that moved in last summer will return. 'And when exactly,' he asks, 'do the swallows come back? I feel they should be here.' Any day now I tell him. 'Soon after the cuckoo.'

Sounds in the rural countryside are the things that don't change. They are part of the fabric of the landscape and our living in it. But in my heart I am dreading the sound of the turbines coming. How will their whirring and whishing and thumping become

*part of the soundscape and fabric of our daily lives? How will
I integrate it? Am I too old to adapt to what is?*

⟁

April outdoes all our efforts to keep up with it. Chris is
outside working now from early morning until I have dinner
cooked in the evening, after which she goes outside 'just to
take a look' at what has been done, and still needs to be, a
twilit surveyance, often with secateurs and glass of wine. I am
with her half the day, doing the heavier, less skilled work.
I would not know what was most urgent, but I know there is
a lot to be done – not least because Chris often says so. On
her knees in a bed, hair in all directions, she will often say she
wishes she was twenty years younger so she had more energy
and could do more. 'Because there's just too much to do.'

To which my usual response has been to say I can do more.

But today, another tack came to me, and I reversed my
thinking and said, 'That's a good thing.'

'What is?'

'It's a good thing there's too much to do in the garden.
A good thing that we will never see the end of it. It will never
be *done*. Think of it. How wonderful is that!'

'You and your optimism,' she said, and shook her head.
'There is still *a lot* to do.'

'Thank God.'

She didn't throw the secateurs at me, just a look.

Papaver Rupifragum

5

May

What has happened?

It has been one of the simple joys of summer that each year the swallows return. They return to the place of their great-grandfathers, and build a nest overhead in the open cabin. There is some inconvenience: in the swooping of their flights in and out as people come and go, often startled by the speed and dash of the birds, and in the little black-and-white bombed hillock of their droppings, but this is nothing to the joy, first when they return, second when you see, suddenly, the nest has appeared, third, when you can see and hear the beaks of the chicks up there, and last, when the family all take to the air. You feel you've been part of something, something that has happened here for a long time, and with a sense of life, of aliveness quietly continuing, despite all that is terrible and dispiriting in the world of mankind. The swallows are a company, and their visit each year something like that of a theatre group of air-acrobats.

This year they came back in May, about three days after Chris, who is more exactly attuned to the clockwork of this place, said, 'The swallows should be here.'

At the end of the month their nest was in place, and in a way, like all those plants that return in the garden without our giving them much thought, we acknowledged the birds with a private gladness, and thought no more about them. They were back and the cycle was once more cycling on. This, it seems to me, is at the core of my satisfaction now. It is one of the central things I've learned about living in a garden, of letting yourself feel part of a cycle. There's profound consolation in it, for it makes even the winter tolerable.

But today, I went over to the cabin to get a fork from the old meal bin that serves as a tool hold-all, and saw the bodies of the five swallow chicks dead on the ground. I looked up; the nest was intact. There was no sign of any damage. I have to admit it was a shocking moment. It made my heart sink. I am soft enough to feel sadness for the chicks' mother and pagan enough to have a presentiment of foreboding.

The small, grey-feathered bodies were warm to the touch. In that instant, taking one in my hand, I had the natural instinct of all mankind, to try and coax life back into it. As gently as I could, I lifted one tiny wing, hoping that the shock of a human hand might startle it and the wing would flutter and the bird recover from the fall. But it did not. Each of the chicks was newly dead on the ground. None of them had any wound or bleeding. I picked them up carefully, very glad that it was me who had come across them.

This too is life, I told myself, taking them away. But what had happened here? This had never happened any other year. Then I remembered that in the past few days the other bird of summer joy, the cuckoo, had been unusually close in the garden. The previous evening she had been singing in the top of the tree just behind the polytunnel. Could she be the culprit? Hardly. Chris wants to believe it's magpies, and is probably right. But I now worry that this moment will end forever the long chain of the swallows coming each year to the cabin. The cycle will be broken.

❧

A red-orange poppy, bright as an Indian sari has opened above a sea of green in the flower bed in front of the house, shocking everything else in sight like some electrifying force. In a collision of colour, it clashes with the late-flowering red tulips. Like the first call of a cuckoo and the first sight of a swallow, the first burst bud of a poppy shouts 'new life' because it heralds summer, rain or shine. Like the poppy shedding its shell it feels as if I also de-cloak.

Elsewhere in the May garden N finally admits the Japanese cherry, once festooned with blossomed layers like the billowing gowns in Claude Monet's Women in the Garden *painting, will no longer give up its white petals to the wind. It is dead. Even our dear friend Martin is saddened.*

*Blue zigzags its way across the top border now, from the tiny-belled flowers of Jacob's ladder (*Polemonium caeruleum*) to forget-me-nots (*Myosotis sylvatica*) and common columbine (*Aquilegia vulgaris*) and bluebells (*Hyacinthoides hispanica*), the last two being somewhat invasive, especially the Spanish bluebells (another one for across to the ditch). A swelling greenness bursts into leaf, indefatigably; the boxwood balls neon with fresh growth.*

In the glasshouse tomatoes are commanding most of the space with some lettuces and salad leaves and carrots, and so too, two out-of-season purple-sprouting broccoli plants that somehow grew inside. Self-seeded? Their yellow flowers floating on the wind up and over the glasshouse and down through the roof windows? A fraction of gardening is pure mystery. The leaves are giant, blue-green, curling and draping gracefully downwards. N decided to transplant one to an outside bed. I told him it'd never work. Another fraction of gardening is pure experiment.

~~~

These heady, blowsy days of May, the idea that there is always more happening below the soil than above it makes your head spin. So much is happening within sight and smell that when I pause in a vegetable bed and think of that, of all the constant, invisible, acts and transacts, six inches below my hands, *and upon which everything depends*, the deepening in me is as though a trowel has penetrated settled thought, freshening it, and with renewed, if not reverence certainly respect, I resume.

We have grown peas here for as long as I can remember. When we first came, before the internet and online ordering, it was impossible to find the kind of sugar snap pea seeds we had grown to love in America. So, each year we would get a Burpee variety posted to us from New York, and for a time had the novelty of the only sugar snaps around, peas so delicious that when Chris was asked how she grew them so sweet she said she put sugar in the soil. And just for a moment, if you had one of the peas in your mouth, you'd believe her.

Well, the sugar snaps have been a staple in Kiltumper ever since. Like all things, some years they are better than others. Last year they were poor enough, most likely because they like dampness at the roots – which for thirty-four years was not a problem hereabouts. But last summer we had the three weeks of drought and when Chris was away at a family funeral in America I probably gave them too little water – '*Niall!*' – and they grew thin and weak-looking. Ah well. That was that year.

Most years the peas grow to ten feet, and although they are lightweight and no shoot thicker than a string, when they are all together entwined they make a significant bulk. So, a good fence is required. In previous years this was often a fairly brutish-looking wall of chicken-wire attached to posts, a wire-wall that often proved too short and required a mid-season, even more brutish-looking, extension with tacked on pieces of wood and an extra three feet of wire.

Now, I should probably make clear here that I am more or less hopeless when it comes to construction. My spatial awareness is poor at best, my sense of the practical maybe skewed by a characteristic novelist's weakness for believing things possible that are in fact not. The thing is, I know this. In a third of a century living in Kiltumper I have had to try and do many things that could be done better by almost anyone, and always by Chris. So be it. Financial reality, I would argue, also goes some way to explain this. We don't have a shed full of tools, and rarely have the 'right' one for the job available. If you saw some of the jobs we have done with the tools we have, you'd be awarding us medals for trying, or naivety. Of course, part of the fault for this is that I would always choose a visit to the bookshop over the hardware shop. Hardware shops are terrifying places. There are thousands of things in there, none of which you have the faintest idea about. At least, I don't.

Well, whatever about that, I would argue that, after our own fashion, we have managed just grand.

The wire fence worked, after a fashion, yes. But there were problems. In the winter, when the gales came sweeping in off the Atlantic, the big fence got caught in the wind and went over to an angle of ten o'clock. In the early spring the fence itself got in the way of digging and renewing the soil with dung, which the peas like. As you dug out, you weakened the grounding of the posts, and the fence, ten feet wide with extensions to ten feet tall, tilted even more and was impossible to set right again no matter how much hammering I tried – me on the top of a step ladder, Himself holding the ladder on the uneven ground, and me saying, 'This won't work no matter how many stones you set in around the posts!' For the six months while nothing was growing, the fence looked brutal. I think the earth was telling us something we haven't paid enough attention to: crop rotation. Peas are great for adding nitrogen to the soil. I make a promise to it that next year I will plant brassicas in the place where the peas grew, as gardening wisdom says brassicas follow legumes.

⤛✕

For the peas of the past two years, we have used bamboo canes set in circles, a kind of teepee, bound at the top with a clay pot on top to keep it together. These had the advantage of being beautiful. Which is an important one for me.

But Chris said the peas weren't happy.

Now, maybe that's only a sentence that a gardener understands: *the peas weren't happy.* And, understanding it, knows you can't argue with it. The cane teepees were beautiful, yes, elegant, all year round, and with a sense of something at least as old as the Native Americans, but the peas weren't happy. They didn't appreciate the space getting smaller as they climbed, the ones on the south side blocked the sun, etc., etc. (Chris tells me the Native Americans probably didn't grow peas. They planted what is known as The Three Sisters: corn, beans and squash.)

Management received complaints, and decided to address them.

So, this year, Chris says, 'I have an idea for the pea fence.'

'What does it require?' I am thinking there'll be a trip to the hardware shop.

'Just long canes and string, but more canes and lots of string.'

It's a small step then to the two of us being outside in the bare raised rectangle of the pea bed that has been deeply dug over and manured, an armful of the longest bamboos I could get at the ready, a ball of green twine and a pair of scissors.

'This mightn't work,' Chris says. It's a common doubt for her, a common impulse after a bright idea, and I've learned to bypass it. She looks at the bed, finds a weed just starting, and another, and picks them off while she's thinking. By way of commentary, there are birds singing in the old sycamores. I'm wondering if they were the same birds that watched us put up the teepees two years ago, and the chicken-wire internment-looking fence the year before. I'm wondering if they know the peas go here and that soon the soil will be opened a little and worms moving, and that the sugar snaps are like sugar in the earth. I can follow that thinking right to the place where those birds *know* us, like their grandfathers did, they recognise and know we are the gardeners of this place and know what plants and flowers and vegetables we try to grow each year, and they and their song are as much a fixture as we are.

'Give me the first one,' Chris says, bringing me back to earth. She places the cane in the right-hand corner of the bed and, as has become a natural way for us, she sets it exactly where it should go and then I add the force to push it in while she keeps it at an angle. Seems right too, I can't elaborate.

We get a second one in the earth on the left-hand side. Then I bring the two together and, standing on a wooden chair, Chris binds the tops of them. We do the same thing four more times, so that soon enough there are five triangles standing, and to be honest, fairly frail-looking on the bed. Next, she takes a cane and runs it along the top, joining the five triangles and holding them together.

'Is that angle the same as the others? They need to all be the same.'

'It's Pythagorean.'

'Now the tricky part. If this works it will be not exactly self-supporting but self-reinforcing. Is that a word?'

'Kiltumper-coinage.'

'If it works.'

'Of course it will.'

'You and your "of course",' she says, but she smiles too. She takes a long cane then and begins the careful business of 'threading' it in and out of the standing bamboos at the bottom of each side. It works beautifully. Each cane, taking pressure from both sides as it is threaded through, locks the whole construction tighter. So that eventually we have a fence that is just a thing of beauty and simplicity, made of nothing but canes and string.

Some peas go in directly, a quarter. (This year we are going to stagger planting and use four varieties.) Then we step away.

But here's the thing. In a way, maybe the most essential way, we are already nourished. That's how it appears to me anyway. Standing there is a thing we have made together, a two-person construction of hope and faith in the summer ahead, *self-reinforcing*. It feeds me just to look at it.

*

A well is a mystic thing.

Growing up in Dublin, I can safely say water was not something I ever thought of. It came from the taps, that was it. When we moved here one of the things we first learned was that the house had a spring well. 'A very good well, Niall,' Mary Breen had told me, pointedly, and in the way she said it I caught if not exactly reverence then certainly respect. A well was an important thing for a house not then, *or since*, connected to any mains water supply. This is not a choice, there is no mains water available here even if we paid for it. Tried that. The group scheme fell through. That's a discussion for another day. In any case, I remember when Mary told us about the well, I conjured up the image of one of those English picture-book wells ringed with bricks, a hat roof, and a winder with a bucket on a rope. None of which, of course, bore any resemblance to our well in the Grove.

Ours is fed by an underground spring, located in the damp far corner of the field east of the house. There, a cracked flagstone lies atop a low little-more-than-a-semicircle of stones, with a large fern growing out one side and a tangle of honeysuckle going every which way. When you bend down to it, you see that the flagstone shelters a shallow – eighteen inches deep – pool of clear water about which midsummer flies buzz and ignore you. You see too, below your knees, how fine is the construction of the stonework, a curve that could have been drawn with a compass, all the more impressive because you can see the stones have not been cut but had to be found and placed once in the long ago.

And it's that that stills you, that's the thing that brings you to that place somewhere between respect and reverence when you're kneeling by the well. When you're there you know you're in a *place*. I know that sounds absurd. But bear with me. What I am trying to get at is the other use of the word, not just a location, but the idea of something that has been, well, *placed*.

⤳

'And, well-placed by my ancestors,' I say.
  'Yes, but also before them,' he says. 'By an ancestor old as earth.'

⤳

Placed, with intention and design, on two levels. First, the unknown and actually unfathomable one, whether by geology or God, your choice, that made the spring rise in this very spot, the placed-ness of that that goes back into the great old age of the land itself. And second, the intention and design that flowed from that, which meant that one day, discovering the spring and gathering the stones to secure and make it serviceable as a well, the Long Breens decided they'd build a house in the next field.

The well, then, is the origin of the place.

And it feels like it too.

When we first arrived, nearby, feeding from the same spring, a pump house had been built to pump water the forty-five metres or so to the house – a task which exhausts the pump every six months more or less and it conks out and suffers the indignity of having its insides taken out on the grass. There's some reinflating and reassembling, the pump refusing to pressurise for a final time until owner and plumber reach the place of complete mystification as to just why it won't work, then it clicks on.

Now, in all the years we have been living here there have been many compliments to our 'beautiful water'. 'That's pure spring water, isn't it?' And for years, every second day, Mary's brother-in-law, Joeso, used to walk back the road with two bright orange plastic buckets to dip them in the well and bring them home for drinking water. (He was doing this when one day we discovered him sitting alongside the ditch, having suffered what, with a kindness in the language, they call here 'a weakness', the two buckets spilled on the road beside him.)

From the beginning, the well has been mysterious. How did it give us water all the thirty-four years? Bathe our children, ourselves, water the garden, allow us to drink copious cups of tea. All these years the water level never rose or dropped. It was just there, in that hidden corner of the field this eye of clear water. It was not for nothing, I thought, that religions appropriated wells, that the first thing Christianity did in Ireland was to take the wells back from pagan rites and rename them for saints. The well is a presence, and for thirty-four years an omnipresent one; that is, it never dried out.

Until last year.

Chris, who has been slowly trying to clear away the brambles and rushes and grass and reclaim a path to and around it, says the well was fighting back, letting us know it has not been happy with our if not neglect then certainly complacency. Maybe so.

Last summer, whether a portent of climate change or not, we experienced what some say was five, some four, dry weeks between June and July. And to my shame — because we grow blind to what in our lives is omnipresent, at least I do – I did not think of checking the well until it was too late. The pump conked out. I went to take a look and discovered that the water level in the well had vanished to leave just a greenish brown mud. The sight of it stopped my heart. I had to take in: *this has happened*, but also: *you let this happen*. And almost instantaneously, I had to battle a feeling of portent. The well had gone dry. It felt terrible.

Getting machinery in to dig for a deep bored well would have been impossible, never mind beyond our budget.

The first thing I realised was there was nothing I could do. Well, I could pray for rain, at the very moment when everyone in the country was delighting in the first proper summer weather in years – all of them with access to mains water. The second thing was that the garden would suffer.

So, a new set of priorities was established.

As in all crises here, neighbours are the way you get through. When our neighbours, Martin and Pauline, built their new house they bored for water, and found it 250 feet underground, so they had water. We borrowed barrels and bins and began filling. Then, as each day came drier than the one before, we began a new regime of carrying water into the house and out to the garden. Each day Chris had to decide which plants were most in need.

༄

Astilbes – those hardy, damp-loving plants, which even tolerate waterlogging – do not *like to be dry. Their leaves and feathery plumes are withering. But an astilbe is tough, and its roots will grab what moisture there is and it will come back next year. And if not, well, garden planning courtesy of the weather.*

Geraniums are so forgiving with this lack of water. The powder baby-pink, double-flowered, ivy-leafed hybrid pelargonium whose name cannot be identified (that I can find – maybe Monty Don knows) graciously accepts any little drop we can spare. But it's the delphiniums and the annuals in the pots in front of the house and the tomatoes and the old box globes that are getting the preferential treatment. They all appear to like left-over dishwater.

༄

It is only in moments such as this that you get to reset, as it were, the way your behaviour has evolved, and you realise with some shock how thoughtlessly wasteful you have been. Now we carry a basin of used dishwater out to a small cedar in a pot. We only boil the two mugs of water you need for two mugs of tea. Only one time do you need to fill twelve litres of water into a cistern to appreciate a flush.

We had no water at all for five weeks.

In this kind of situation, Chris has a genius. In our earliest years here she had described our life to her brothers and sister in America as 'a kind of camp-out'. And maybe because it seems we are always inventing the way we are living, because, like the garden all around us, life here is a constantly evolving thing, she adapted to the dry well and somehow kept the gardens going.

When eventually it rained, I visited the well daily, and learned the lesson that a downpour does not replenish what weeks of dry weather has dried out. Not two or three downpours. Not rain showers on-and-off for a week. Slowly, slowly, the water returned. But when it did, when I went out one morning in early August and saw that the level was back up to the moss ring, it felt, well, like a blessing.

But what is that Eliot line about having the experience but missing the meaning? There was no point in going through the drought unless you learned something from it. I do think we have been much more watchful of water usage since. But also, even though the general opinion was that last year's heatwave was a once-in-thirty-years event, as we approached summer this year, we decided that it was just foolish not to make more use of all the rain that falls on us. So, I set up a water butt at the eastern gable, another at the western one with a tap on it for Chris because dipping and carrying is not advisable for an osteoporotic frame. The butts filled in a day. Besides the wonder of seeing a measure of just how much rainfall there was during the night, the butts reversed the way I have thought of rain. Now, I look out the window when it is pouring down and think: *Great, the butts will be brimming.* And soon enough realised we need many more. Whether or not the summer to come brings the same kind of drought, I have to say it is deeply satisfying to dip the watering cans and bring the rainwater to the plants. It's a simple thing, but it makes me happy.

The pump is happy not to be used so much too.

*

A moment of dizziness and sudden heat when you lift your head from weeding and the air spins and it seems you will fall down in the garden before you can cry out. Chris is absorbed over by the tunnel and could not hear anyway. There is no one to hear in the green serenity, and on the very point of losing consciousness your mind flashes through a condensed version of a number of scenes beginning with your falling sideways onto the grass, the surprise of that, that you are actually falling and fading at the same moment, in a heartbeat you will land on the ground, your form lying there unconscious, the shock for Chris however much later when she comes with secateurs thinking at first that you have chosen to lie down and in the same instant knowing this is not rest. And what will she do then? The whole thing flashes past, and somewhere inside of you, because of the shock of those scenes perhaps, because some part shouting *No!* is refusing this version, you blink and breathe and steady, and in a breath, another, and another, the moment passes. You don't faint.

You might think that, working in the garden, you are as 'grounded' as possible, but still it happens, happens to both of us now, and is just another of the tells that we have lived a life inside this garden, and may one day fall in it, no different to the petals of each rose or poppy.

Only not today.

In his excellent book The Year in Ireland, *Kevin Danaher writes about* Bealtaine, *the festival to celebrate the first of May, the beginning of summer. The name is derived from an old Irish word meaning bright fire. Once, bonfires were lit in celebration. May was the time of year when women gathered the May dew, believing in its virtues as a medicine and a beauty aid. 'The dew was believed to bring immunity from freckles, sunburn, chapping and wrinkles during the coming year.' It may even cure or prevent headaches, skin issues and sore eyes. You should go out into the early morning to the meadow and knock dew off the grass with your hands. Collect it in a jar or on a plate. Let it rest in the sun and once the 'dregs and dirt' have settled pour off the good stuff until you have clear but whitish dew. Apply it to your face. Walking barefoot in the dew is a cure-all for everything, including all sorts of foot ailments. Better still, he asserts, be assured of good health and a fair complexion for the coming year by rolling naked in the morning meadow.*

'Niall?'

6

June

The work on the wind turbine development has begun in earnest now. The clack of diggers on the hill wakes us in the early morning, and when you walk outside in the garden with your first tea you're instantly aware of the noise. As in all things, you only realise how quiet this place has been all these years when it's not quiet any more. You hear the birds and the breeze sounds in the sycamore and ash, the breathing of the garden, and then *clack-clack*, a harsh metallic encounter with rock, *clack-clack* again, as the diggers claw the hill of Tumper, and now you are in the business of trying *not* to hear them.

Since we lost our appeal to the planning authorities, we've known this was coming for some years now, but now the reality is here, and whatever of the theories and arguments in favour of wind development, when you're living alongside one that reality is not green, elegant or without cost in earth and human terms.

Nonetheless, that argument has been won by the developer, and it seems we live in an area that has been designated not of scenic value – a surprise to us, and I have to admit a little hurtful – but perfect for wind development. Consider you've spent thirty-four years trying each day to contribute to the beauty of where you live, trying to be a good enough neighbour, and are one morning told by someone who has never stood in your garden or walked the road that it is not beautiful enough not to be exploited by turbines. The wind industry has moved quickly, and there are already several turbines spinning on the skyline to the north of the village, and more due. The sloped back of Clare's most significant mountain, Mount Callan, is whitely stippled with them and always makes me think now of a bull, wearing the matador's lances and the little flags of the *banderillas* not yet fallen, but its dignity destroyed.

But this is just me and Chris. We are perhaps alone in despising the clever abuse of language that is the term *wind farm*, for who could be against a 'farm'?

Of course, in my view, it is not farming, makes a mockery of actual farming which is about caring for, nourishing, supporting, improving, minding, husbanding a portion of the earth for the next generation coming after you. Farming is not about profit per se. It is about living on and in and in some ways *for* the land. And for true farmers, in the decades we have lived in Clare, we have learned a deep respect. They know their fields, from long living together know the good and the bad of them, and anyone who has seen an actual farmer come in after a long day out in the fields knows that farming is hard work. In no future time can I imagine a developer coming in, toeing off his wellingtons and saying 'hard day's wind-farming'. What I picture instead is someone at home, a home not anywhere near the noise and immense shadow flicker of the actual turbines, flipping open a laptop and checking the figures for today. A 'farming' entirely removed from the land. A wind development is not a farm, except in the very narrowest sense, but it is a clever theft of language.

But again, this is just me and Chris. I am old enough now, and out-of-time enough to be a crank. So be it. I am in favour of wind turbines, only *not* where people are living beneath them, *or might in the future be living*. So, not on land, of which no more is being made.

Now, to be clear, I bear no personal animosity to the developer, he is just doing what he can to maximise profit from what they call an 'outside farm'. As with the case of the vast majority of developers, he does not live beside the turbines. Neither do any of the politicians who have approved the policy, nor do the planners, who at their desks in Dublin have decided that while the same turbine that in Germany has to be 1.5 km from a dwelling, and in Denmark 2 km, in Ireland can be 500 metres from a house.

In this case, ours.

In Kiltumper, what the 'green' industry of wind-farming looks like at close hand is lorries and diggers passing from early morning. It is a signpost at the end of your road saying, 'Deliveries to Kiltumper Wind Farm'. It is portable toilets being delivered. It is the 200-year-old stone wall alongside the road you walk every evening being taken away in half an hour to make room for the turbines coming. It is the fields to the immediate west of you, just across a boundary stone wall from *Pairc na moine*, transformed as hundreds of metres of topsoil are ripped away to make roads up to the site, and then as many loads of gravel being delivered by diesel lorries, the gravel being pounded down and levelled by diggers before the next load comes. It is all day. Wind coming as it usually does from the west, if you stand in our back meadow the air is diesel-ed and edged with engine noise. In a place like this, where in all the years we have lived here, change has been minimal, where Lena's farm, as we called it, was a half-dozen ragged fields home only to a herd of placid cattle and flights of birds, any activity was always going to be impactful. But in a few days the transformation of the old farm is startling.

The works going on behind us won't be forever, but they will be breaking stones and making roads, excavating the giant craters needed for the bases of the turbines for the rest of the summer. So, there is that to accommodate now. Although we don't see it from the front garden, because it is only 500 metres away the work is hard to ignore, and when you walk past the house along the gravel towards the glasshouse what you see on the hill of Upper Tumper are the outreaching orange arms of the giant crane.

It mightn't be so distressing if Chris didn't spend five or six hours every day out in the hush of the garden, which is actually her way of life, our way of living here. And it mightn't be so impactful if I wasn't a writer, who each day depends on that quietness in order to work.

None of which of course *matters* to developers or government. We are the small picture, not the big one.

When the turbines are up and spinning, we may get used to them. 'Think positively', someone who lives in Dublin advises us on the phone. He is not as often outside, commutes to an office by car, comes home in the evenings.

Well, I think it took us thirty years to get used to what Robert Macfarlane calls the 'pylon's lyric crackle' from the power lines that cross the land. And we're still a long way from thinking it's lyrical.

We did eventually get used to the pylon crackle, which occurs when the falling rain is mostly misting. The power line sings. But we knew it wouldn't last forever. It comes and goes, and in truth is not that often. Because usually here it's not misting, it's out-and-out properly raining. What we were first fearful of was the magnetic field from the high-voltage lines. The Electricity Supply Board came to measure it for us and said we were safe and far enough away – as long as we weren't standing under it, in our field, *for long periods. 'And if we want to stand in our field for long periods?'*

'You can, but I wouldn't.'

We have lived thirty-four years with it, in the early years raised horses and cattle to no obvious injurious effect. But the noise from the turbine blades will be an entirely different affair. And they won't be turning just when it's misting. They will be turning constantly.

A Google search is not encouraging. The EPA Ireland states: 'Traditionally, wind turbines tend to be installed in areas of low background noise and, particularly at night, it is often the case that there may be no other significant sources of noise other than the noise of the wind as it blows through trees and other foliage. Additionally, people often chose to live in rural areas specifically because they value the lack of noise from man-made sources, *so any such noise which is audible may be a source of disturbance/annoyance and possibly complaint.'*

It's this last statement, that people often chose to live in rural areas specifically because they value the lack of noise from man-made sources … *For instance, what if you're self-employed and work from home? What if you're a writer and your workspace is your environment? What if your office is your home, and what if your health and well-being depend upon the garden?*

'Possibly complaint' *is fairly dismissive, I think.*

The WHO published a paper at the end of 2018 stating:

'*The wind industry has denied and ignored evidence directly linking wind turbines and sleep disruption leading to negative human and animal impacts worldwide. Expect WHO's new guidelines to give rise to new standards to mitigate if not eliminate this ongoing suffering.*

'*The burden of environmental noise with wind turbines is not episodic or random: for the most part its effects are constant and unrelenting ... This is an undeniable health pressure of enormous magnitude.*'

The World Health Organization's guidelines advise noise levels be below 45dB Lden. By the middle of 2020 we will know what these levels are and how the 'Wind Turbine Syndrome' is affecting our health, but until then we wonder daily about how these giant machines, set to tower above us up on the hill, will impact the way we live in Kiltumper.

It is a fact: They Make Noise. Unrelenting. Undeniable. Enormous magnitude.

Today, I noticed that the delivery lorries have been driving along the edge of the grass bank that runs by the stone wall out at the front of the house. The face of the bank has become mud and each day a little more verge has been disappearing. *Does this matter?* It is only two feet of ground. From the front room where I write each morning, I can see the lorry drivers, often on their phones, racing past. I know they are used to delivering to an industrial site, not along a country road, and are not intentionally taking away two feet of our grassy bank. Still, I want intention. *We live here*, might be what I want to paint on a sign outside, but settle for bringing down some large stones in the barrow and placing them on the edge of the bank. Hold your ground is the theme.

None of this is what you want to be doing or thinking. None of this is in my nature really, how we have been living or want to be living. It is the simple wish of most people, I think, to be allowed to carry on quietly as they were. In my notebook, this:

Q: What do I long for most?
A: Peace.

And that is what is lost at the moment.

I know that all this affects Chris even more than me. The garden is her church, retreat centre, meditation place, gym, gallery, larder and pharmacy. So, when I hear the diggers clacking extra loud to break into the rock of the hill, I know she will soon be distressed and I find myself suddenly talking loudly to try and mask it.

'What do you call that again?'

'*Coreopsis*. Why?'

'No reason, just, looks well, doesn't it?'

She plays along knowing full well what I'm doing. What else can we do?

Clack. Squeal. Hammer.

*

A different class of disturbance today.

In the middle of the day, while Chris and I are tying up poppies, four large black American helicopters fly in formation across the sky just above us.

They are bringing the American President to his hotel in Doonbeg, nearby.

The beach we walk is closed to everyone for the next three days, the village has Garda roadblocks. It's a surreal moment, hearing the great roar of their engines over Kiltumper and watching them almost in slow motion cross the pale sky like dark insects.

We don't wave, we deadhead the poppies.

*

For my birthday, Chris gives me a framed copy of one of the first stories I had published. It was called 'Apples in October' and appeared in the New Irish Writing page of the *Irish Press* newspaper. She had found the story while in the midst of one of her 'clearings' – an impulse that arrives every so often, and which I know enough to stay clear of. I had thought the story lost and was startled to see it again.

Two things struck me about it. The first was that it was a story about two older people – modelled I would guess on my grandparents – who were waiting for their son to call to help them pick all the apples off their tree before they fell, and in remembering this, I realised that this man and woman were little older than we are now. It struck me that it was a story about an old couple and a garden, as though seeded in my imagination all those years ago had been something of the life that was to come. And in this was the second thing: the story was published forty years ago when I was twenty-one. I could remember the Saturday the story was published. I could remember going across the Lower Kilmacud Road to the shops to buy the newspaper, not saying anything to Don behind the counter. Could remember carrying it home unopened, not telling my parents or my brothers and sister. Taking the newspaper up to my small room, putting it on the table that butted the bed. An unopened newspaper has such promise folded within it. Something so civil and proper, I thought, buying a newspaper for myself for the first time.

Not for my father or the family. This one, for me. With my story in there, somewhere. If I dared to look.

I didn't, dare or look. I left the paper with its one perfect fold, its creaseless serenity and promise, there on the bedside table, as though opening the pages and finding the story, reading it, could only diminish what was still at that moment in the precinct of the marvellous. Opening and reading it would bring out into the open as it were what was still only between me and the page. The story was still too *personal* to me and I was not prepared in any way for the public experience that is in *published*.

What I remember is that I didn't show the story or tell anyone about it. My Uncle Jim I think discovered it and told my father, and when I was asked 'Did you have something in the paper on Saturday?' I said 'Yes' only, adopting a cooler persona than I had, and shrugging it off, as though the personal and intimate writing in the story belonged to some other fellow.

As, in a way, it did.

Now, thinking of this, forty years later, I see in it something of the path I have followed ever since. In those days, when setting out, I never wished for a literary life, a term that seems to me an oxymoron. I never read an extraordinary novel and thought: *I'd like to know more about that writer.* I thought: *I'd like to read more by that writer.* It was the book that mattered, not the writer. For in some way too profound for me to explain, the book was more personal, more intimate than the writer could be. The book was the thing. In this I am with those who have said it does not matter who Shakespeare was, it matters that *Hamlet* is.

None of which is very helpful in an age when personality is king and 'followers' are counted. But so it is. And when

I looked at the framed newspaper page today, the fact of the forty years on the path of trying to write, the fact of *still trying* was moving to me, knowing that on that path, while writing, I have experienced some of the greatest joy and oneness with the world that I have known. Standing there, I passed a silent nod to the 21-year-old placing the folded newspaper on his bedside table.

'Happy birthday,' I said.

*

Seven years and one day after my story about my grandparents was published, Chris's first story, 'Queens Irish', was published in the same newspaper. It was about her grandparents.

*

Thankfully, in June, there is so much happening in the garden that it takes all your attention, the urgency of things rushing now to bloom. How can it be that every year it is surprising? The garden writer Jim Nollman has called it Mother Nature mating with Father Time, and it's true there is an erotic frisson in the air, even in old Kiltumper. Every sense is engaged.

Along the eastern bed the roses are come into their own. I'm not exactly sure how it happened, but the rose bed is, nominally at least, under my care. It may be that, overwhelmed by the very large number of different plants in the garden and how daunting it seemed to learn the nature of each, out of some familiarity I gravitated towards the rose. It may be just that the rose is the flower you most encounter in literature, and almost all the most engaging experiences of my early life and youth came first through literature. Or it may be the influence of Granda Williams. As I've said, the house in Dublin that I grew up in had a small front garden, but down one side between the concrete and the obligatory patch of lawn was a narrow bed of roses. (Now, it wasn't a proper garden in some senses, it was small, the ground a skin of earth over builders' rubble, and a hedge grown not for beauty but privacy. But let me say here that I have come to understand the noun 'garden' has the magical ability to be both the enormous, precisely cultivated, grounds of estates, as well as the six-by-three-feet balcony of our daughter Deirdre's 'garden' in Brooklyn. A garden is anywhere something grows under care, is my own definition.)

Now, my father was not a gardener as such, but his father, who had been in the First World War, captured, and thought dead, one day came home smiling, had a green thumb, and a particular fondness as it happened for roses. I can't say why. When I was small, he and Granny came to our house every Tuesday. When it was raining, he might watch the wrestling or the races on the television. But often as not, he'd go outside and 'take a look at the garden'. During the summer, he brought with him in a brown paper bag what looked to my boyhood eyes like something he'd had with him in the war. It was an old brass pump with nozzle attachment, and when he had made up some compound in the garage he'd go out and shoot it at the roses. A big man, without a lot of talk, Granda contained within him the mystery of all grandfathers, whose pasts are too large to be known, and the fragments of which – war, imprisonment, horse races, wrestling, roses – combined in a mystique that was profound and compelling. Often, I stood behind him as he shot the roses with a swift pump action, shooting them and thinking of I don't know what, but it's not a far reach to suppose what he'd seen in the trenches in France never left him. In any case, he sprayed with vigour, the stuff misting out over the roses – and often enough over both of us too, no face masks or precautions in those days, and when you went to bed that night you might be aware of a chemical scent that had escaped your two-second slapdash handwash in your pyjamas, but what of it, it was the smell of your grandfather and roses. For pruning he'd draw from his pocket a two-bladed penknife; the cream enamel of the handle was embrowned and cracked, but he didn't want a new one. (We had his British army bayonet knife upstairs, a thing your boy eyes could look at for a long time and imagine its many stories.) He wore no gardening gloves, his skin immune to the thorns that always stuck into you ('Granda, I'm bleeding!').

From the distance of beginning to approach his age now, I realise that my father asking his father to come and 'look after the garden' was most likely a ploy. The garden was too small to need much gardening as such, and instead, it was a kind of bridgework between two men who were not talkative or demonstrative, but loved each other. After my grandfather died, my father kept up the roses. Some summer Saturday mornings, after he had washed the car, and before he went to golf, wearing gloves and using secateurs, he'd 'tidy up the roses'. So it was that in our early years in Kiltumper, when he and my mother came down on one of their few visits, and he'd step outside into what was then a jungle of untamed 'garden', his eye would look for and find the familiarity of a rose. *'Could do with a tidy-up, Niall,'* I could hear him saying, though he didn't allow himself to say it. In his last year, after my mother died and he was in a nursing home in Dublin, I was the one driving across the country to go see him. One of the memories of my life is of taking him out in the wheelchair, getting him into the car to go and visit the old house a last time before it was sold. We drove in and parked, and before

I had got out to go around and help him into the chair once more, he said, 'Just wait a minute.'

We sat there in the car alongside the rose bed. I think I knew enough not to speak. He just looked at the house and the garden. He did not weep, or show the slightest outward sign of what he was feeling. But you knew. It was like standing at the edge of a very deep pool. In the middle of the Dublin afternoon, we sat there in the car, and looked at the house and the roses alongside us.

All of which may, or may not, go some way to explaining how the roses became my jurisdiction. Perhaps all gardeners have this secret underground root system, familial, historic, associative, or imaginative, connecting them to whichever plants they most gravitate towards. Known or unknown, I suppose. Why does one person so love dahlias, and collect dozens of them, and another geraniums or primroses say? Chris loves poppies. Do individual species of plant 'speak' to certain individuals? It's a head-scratcher, as they say, another garden mystery.

Of course, an alternative explanation of our Kiltumper arrangement may be just that, with a wife's genius, Chris wanted to engage me more personally in the upkeep of a part of the garden, and thought I would do least damage in the rose bed. You wouldn't put it past her, and she does secretly oversee everything anyway, dropping the occasional prompt, gentle, or not, if she sees the first black spot.

In any case, although the truth is that I have no expertise, I do have a love for the roses, and when they bloom as now in June it lifts my heart more than any other element of the garden.

Garden tasks are plentiful for the Chief Groundsman who graciously accepts the chores I ask of him – hedging, mowing, clipping, transplanting, barrowing, carrying stones, cutting brambles, raking, cleaning the gutters, digging, spreading bark mulch, edging the lawn, weeding the big stuff like docks and nettles and Japanese knotweed. Today he was on his knees cutting away the grass from the flag path on the way down to the lower cabins. The place we call The Shebeen. I leave him to it. Hashtag: gratitude.

One of the old-fashioned roses, which Chris tells me is either 'Falstaff' or 'Munstead Wood' of the David Austin series, is a densely packed heart of cerise petals with a pure perfect scent, exactly what I think of when I think: *rose.* For no reason other than wonder, with utmost care, I take apart a fallen bloom today to count the petals. As I count past ninety, I am already thinking *It cannot be, it can't be that there are exactly…* But it is true. There are exactly one hundred petals. Honest to God. And standing in the garden I go through thinking this must be just outrageous chance – *Should I pluck another and take it apart to see?* How odd might that be? Man in a garden in west Clare taking apart roses? And what if that one also had a hundred densely folded petals? Would I need to pluck another to be sure? And then a different rose? And so on. And in that same moment I allow myself to imagine the hundred petals in the bloom of a rose as a kind of numeric fundament, an origin story maybe for the decimal system in Arabia. *Why not?* I am not the first nor last to ask just how many petals make a rose, or to consider that in an instance of mathematical perfection is an order that is, well, divine. In the garden today, I hold the deconstructed flower in my hands and breathe in the last scent of it, and all thought flies away in the sensual enwrap of that perfume.

Each of their scents is so particularly individual it seems astonishing that there can be such individuation, such variety in the thing of a single name. And the bed only has a dozen rose plants. I am not sure about the yellow one, whether I like it or not, but Chris assures me it's the most consistent bloomer and our oldest rose. It is in a bed of mostly pinks and reds and a white given to Chris for her sixtieth birthday and in honour of the publication of her first novel, *Her Name is Rose*. And the single dusky 'Blue Moon', which, once you smell it, just an incredible mix of vanilla and honey, you think: how could you not have this?

*

Poppies, giant orange-red poppies, blooms bigger than teacups, are one of the most characteristic features of the Kiltumper garden in midsummer. I think Chris's special love of poppies probably came from *The Wizard of Oz*, and we do have some of those small scarlet ones that feature in the film. But there are also orange California poppies that self-seed in colonies all along the paths, and are allowed to stay until September. One time we did manage to grow a single blue one, and this year have a strikingly cerise variety, as well as a number of pale salmony pinks which are the first thing you see when you step in through the open cabin. So yes, poppies. June is their month. And across the vista of the top beds your eye lands on them and carries you along. The vivid lush invitation of their colour along with the tissue texture of the flowers is – Jim Nollman is right – a June erotic.

Colour and scent and shape and texture. The invitations of June are many.

*

It strikes me that the first purpose of these notes is simply to say this is not, in Robert Macfarlane's phrase, a 'nothing-place'.

There is no such thing on the planet as a nothing-place.

As I've noted, he also writes that a landscape that is undescribed becomes unregarded. I especially like this, for it encapsulates my own sense of purpose in writing. In the time we are living now, when it seems the countryside is increasingly being disregarded except for what resources – water, wind, people – it can offer the city, the act of looking at and describing one particular place is not only an act of preservation, a walling-around with words of something you feel is in danger of disappearing, but more, it is to acknowledge the richness and beauty of that place, and, in an antique phrase, to give it, quite literally, its due regard.

Or something like that.

I can't quite catch it, but I have this core feeling that something in this is right.

I am aware that some of my sense of urgency to write about Kiltumper this year stems from the decision of the planners to designate this a place suitable for wind exploitation. Besides all the other objections we have to this, besides knowing that no planner stood in our garden, or on the hill of Upper Tumper, in fact did no actual regarding before making this decision, there is simply a feeling of hurt on behalf of a place we love.

Is that absurd?

It is the truth. If you can love a place, as we do, it follows that you can feel hurt for it too. And the only remedy for this, for Chris, is to spend more time inside the hedge line of the garden, literally her preserve, and for me to try and hedge-in this place and the life we are living in with words, the purpose of each of which is to say, Look. This, like all country places, is a special one, rich with human history and nature. Look!

*

Soil. Sun. Water. Three things you learn to appreciate fully in a garden.

The peas are up and starting to climb Pythagoras's Fence. Because of the distance between the horizonal support canes, we always knew they'd need some extra help. So, this month, every week or so, Chris goes outside and 'knits' a line of green twine in and out of the uprights, making not so much a pea-fence as a pea-jumper.

The peas seem to like it.

*

Today we go to Chris's oncology appointment in Galway. These have been a feature on the landscape of the past four years since she collapsed in London and had a large tumour removed from her bowel. None of the time since then, the recovery from the surgery, the chemotherapy that she became allergic to, the aftermath of that, the vast solitude of all serious illness, has been easy. Fighting is the word they use for cancer, and time and again, you realise how apt that is. Nor does the fight quite end, and there is a kind of ongoing spiritual and psychological impact in that. Nonetheless, the cancer patient's treatment is measured in oncology appointments, and the further you get from the surgery and the chemotherapy the further apart the appointments become, so you have a sense of moving away and gradually, if you are very lucky, into a new terrain. This is how it has seemed to me. Through all of it, intense and against extraordinary odds, Chris has battled – the adverb that has occurred to me is the old one – valiantly. And whatever help I and our children have been in being by her side, near or far, the real support troops were certainly in the garden in Kiltumper.

I'm not sure how much has been generally written about the healing power of working in a garden, and whether such things are dismissed by the medical community as belonging to the airy-fairy. Well, in Chris's case the facts I think are indisputable. Getting outside the way she has, even during the wretchedness of the chemotherapy when her body was thin and weak, going about the flower beds, trying to do exactly the same work she had always done, only this time with me close by to lift or cut back or follow whatever General's instructions came, has meant a continuity, a carrying on living, because that is one of the prime lessons any garden teaches you: the garden grows on. Just that. Everyone who works in a garden knows they are tending to some things that will outlive them. And of course, when you are dealing with serious illness there is a sharpened edge in that. Still, that is the covenant, and it is immensely moving to me. Some late afternoons I would leave Chris and go inside to make the dinner, and when I'd come back to the window to see that she was all right I'd see the girl of wonder she always has been bending to help pick up a weighed-down blossom or pulling some strands of couch grass that had come up near the base of a peony.

At first, Chris's appointments at Oncology were spaced like seasons, every three months. Eventually, they moved out to every six months. The ritual remained the same, blood tests and scans in the lead-up, then the trip to the university hospital in Galway, the search for non-existent parking, the queue for the check-in, the wait in the general waiting room with a coloured slip of paper, the announcement on a PA system out of the last century coming too blurred to be understood but a general shuffling among the hundred patients and a consensus emerging that the Pink Tickets had been called, *Yes, 'twas the pink ones, I think,* up the two flights of stairs the laboured traffic of some very ill people to the waiting room for Oncology, a room with twenty-eight seats, all of which were already filled, a number of men and women standing along the corridor, leaning in to shoulder the wall, or stepping into doorframes when a wheelchair or file-trolley came past, the 'weigh-in' (*Fighting cancer, patients like boxers,* is in my notebook, *only here weight-loss is one of the enemies. When C comes back from the weigh-in with no weight lost she gives me the two thumbs up, and in that moment echoes the gesture in his last days of her grand-uncle Mike McTigue, featherweight boxing champion of the world, 1923–5*), and then the wait. There is no wait quite like the wait to see the oncologist. You can try and talk through it to distract each other, both knowing that that is what you are trying to do, you can look at your phone, read a book, or watch the muted mid-afternoon television overhead. But really, none of these helps. You're in a pale-yellow space, painted walls, institutional polished floors, posters for cancer counselling and where you can buy wigs or treatments for hair loss. There is absolutely nothing green in sight, so it feels to me we have left the world proper, certainly the one in Kiltumper we live in. This is a kind of halfway place, a sterilised limbo you've entered. Now, you're waiting for the verdict, same as everyone else in that corridor and in that waiting room, many of whom have come long distances, from farms in Sligo and Mayo, for the ten- or fifteen-minute appointment that will tell them how they are doing in the fight.

In Chris's case, over the years, we have had all manner of encounters with a large range of doctors, some junior, some on their oncology rota. Her bloods have been 'lost'; her file – it is a fat and well-worn Manila folder, like an unruly garden, with a rubber band keeping in place the multitude of pink and blue and yellow and white sheets that have accumulated and multiplied in there over the years – has been lost and found; while opening her file, one female doctor asked, 'So, how is your breast cancer doing?' 'I have *breast* cancer now?'; so we have come to expect the unexpected.

Today, the unexpected was to be called soon after we arrived, and be brought in to see the head man, Dr Leonard, who was in charge of Chris's case but whom we hadn't seen in person for a couple of visits.

'We haven't seen you in a while,' Chris said.

'No, generally, when patients are doing well, I don't get to see them.' It was a half second before he realised what he had said and, seeing Chris's face drop, added, 'Not that you are not doing well. You are. You're doing very well. Very well.'

***

My tumour marker is below 2. Dr Leonard shook my hand when I left his office and told me to get ready to have a party at the end of the year. Your fifth-year anniversary, he said. OK, I said, I will, but as we walked to the car I was not as jubilant as I thought I should be.

There are so many emotions it's difficult to explain.

If I put down the bare facts of it, I think I might get a handle on it.

In late January of 2015 on a weekend away to visit our son in London I ended up in A&E at five in the morning via taxi from Primrose Hill into the Royal Free in Hampstead. Luckily N got on the next flight from Shannon and was there by the end of the day. A few days later I was operated on. A large tumour and ten inches of colon were removed, and on World Cancer Day – 4 February – I found out I had cancer. Stage 3b. Good timing for a storyteller. You couldn't make it up. Sometime I will write a short piece about my experience through the Irish healthcare system, an experience which was unique to me, me alone and I'm not saying it happens to everyone. Computers at the hospital on my third session going on the blink and so no chemo being dispensed. Go home. On another day waiting seven hours for my chemotherapy to start. On another, the scanners being broken and N having to drive with me from the Emergency Room in UHG to the hospital on the other side of Galway to get scanned for a possible pulmonary embolism – me with a cannula still taped to my arm. 'You'll be quicker if you drive there yourself.' Becoming so allergic to chemotherapy that the head of Dermatology, whom I waited five hours to see, all the while the allergy getting worse until it looked like leprosy, had to call in his team to show them what a severe allergic reaction to Fluorouracil looked like. 'Look everyone, how's this for an extreme adverse reaction.' From the tip of my crown through my face, lips, nose, eyelids, chest, arms, hands, legs, to the ends of my toes all blistered like smallpox. I had to be immediately covered head to toe in paraffin. 'If you aren't, it will be this, not the cancer that will kill you.' I was a unique case and was advised that I had to stop chemotherapy completely. Of the twelve sessions I was told were vital I got only four. A bit scary as I realised that was the end. No more treatment. Though for me the chemotherapy was truly truly dreadful, I felt robbed in some way. I was on my own.

I think I'm meant to do something with these details… Maybe it's just for me to realise I am a survivor.

We will see what the end of the year brings when I become a statistic one way or the other. I will have the annual CT scan and more blood tests and with continued luck I will be shown my walking papers. Maybe Dr Leonard will see me again and say 'You're on the other side.' But today I am afraid to think about it. I stayed quiet on the drive back to west Clare and then arrived back to the garden, where there is safety.

Perhaps in recognition of the good news, because inside you the relief is enormous and wants to find signification, echo, in the outside world, I ask Chris to walk around the garden with me this evening and to catalogue all that is in bloom inside our hedge line this June.

'But we know what's in bloom, Niall.'

'I know we do. But I don't know all the names. You do. So, just humour me.'

We begin at the lower bed on the left-hand side and move anticlockwise around the garden.

A purple baptisia.

In front of it, looking a little like an extravagantly showy thistle, one shade of purple on the outside of the star flower and another inside it, a centuria. You can't see it from the house exactly, you have to come down close and then you can't believe you could have missed it.

A pink peony of the pale almost whiteish pink.

Twelve roses in full bloom. (Two not yet.) None the same. Mostly pinks and reds, one yellow, two white. All but two with head-exploding perfumes.

Two tree peonies. ('They're not happy where they are,' Chris says.)

More poppies than can be counted. Most with blooms as large as your hand, many an orange-red colour, but also some delicately pink ones.

A blue delphinium that is a kind of superstar in June. It gets all the supports it needs and then grows taller than the tall canes. The blue of the blue delphinium is just extraordinary, and of course makes me think of Delphi and Greece and Grecian skies and the sheer sensual delight of my first swims in the Aegean last summer after sixty years on the planet. (*Can a flower transport you?* I write in my notebook. *Yes.* Every time I look at the blue delphinium I am in Greece in west Clare.)

A fringe of lady's mantle, blooms of palest green, so small as to almost escape what we think of as flowers, until you touch them.

A half-dozen eye-catching pink phlox (*1950s lipstick-pink*).

A deep dark purple astrantia.

A multitude of candelabra primulas. Four levels of pink flowers at equal distance on each perfectly straight-up stem. (*You look at them and think: Who designed that?*)

The miniature daisy-like feverfew.

Three purple iris, that are deep and dark flowers when they first appear and grow paler as the petals fall further open and expose a white centre.

A tower of pink, purple and purple-red sweet peas, that I am quite proud of. Chris grew them in a pot and I placed the pot inside one of those iron obelisk towers outside the window where I write so that the flowers would grow up into view. ('But they will grow taller than that.' Chris said. 'What then?')

A feathered pink astilbe that if allowed a free run will, with its coy demeanour, colonise the bed.

Geraniums. There are so many kinds they could have their own chapter. They are the most common answer to the question visitors to the garden ask: 'What is that plant?' Through the beds there are various winding freeform flows of them, miniature pinks and purples and a purple that is almost blue. They are the flower that most carries your eye, taking you in colour through the broad view until you are stopped by something more obviously striking. *But take your time and actually look at them*, I tell myself. They are delightful. A favourite: the 'Rozanne' blue.

Scabiosa, pale purple flowers floating.

A large purple lupin – the superstar on the right-hand side.

A purple delphinium that can't quite compete with its blue sister across the way.

Another purple baptisia.

Achillea, yellow flowers and grey-green leaves, a subtle combination. Again: *Who would come up with that?*

White Jacob's ladder. (Every time Chris mentions the name, I think I must look up the story, and eventually I do, and after reading of the ladder of prayers rising from earth to heaven look at the flower differently. I can't be succinct enough in saying why, but there is something to thinking of the origins of flower and plant names, the ones that have descriptive elements are one thing, but the ones that have story embedded, as it were, are another, and always connect me in some way to the first set of eyes that really looked at the plant, and then named it. And in that connection across time and gardens is a deep part of the pleasure of plants for me.)

A white rose, on its own at the front, the first thing you see in June when you step in through the stone doorway. Large perfect blooms this year that are side-by-side with those of a light-pink poppy.

Cerinthe, blueish purple, I say to Chris when she says it's blue.

Spires of catmint. Which is purple. ('Why is there so much purple in the garden?' 'A lot of what you're calling purple is in fact blue.')

A white-and-pink rose, nineteen ball-flowers of exquisite blush-pink coming through the white. A stunner. The parent of this rose grew rambling all over a house we once rented in France; not only were the blossoms remarkable, the scent was too. Chris took a three-inch cutting, put it in a Ziploc bag and brought it home. That was many years ago now. The rose has grown well – despite: '*It's in the wrong place, Niall*' – but sadly, it left its perfume in France. In Kiltumper, it has no scent, except in our memory of it. Marine, a young French horticulturist who has moved into the parish, visited the garden and told us the rose was called 'Pierre de Ronsard', a French Renaissance poet whose love sonnets I remember studying. He has a melancholic sonnet called 'Comme on voit sur la branche' about looking at a rose in May.

> As one sees, on its branch, in May, the rose
> In its bright youth and new astonishment
> Making the heavens jealous of its tint
> When, touched with dew, in the first light it glows,
> Then, in its petal, grace and love repose.

(I'm not certain ours is a 'Pierre de Ronsard'. It may be a slightly paler version, or Pierre lost some of the pink in his cheeks when he landed in the rain of west Clare. It doesn't really matter, does it? It made me look up the poem and remember, and associations are half of the secret pleasures of flowers and plants anyway.)

Orange poppies. A multitude. They prefer growing in the pebbles of the walks, and we let them. Walk around them.

Purple spikes of Russian sage.

Foxgloves. Mostly purple. (Chris keeps trying to cultivate the white ones. We have one shy one this year around the back.)

Five more roses.

Flame-orange flowers of calendula.

California poppies, white, and white-pink.

Wild verbena with its tall reddish stalk and soft cloud of white flowers.

Dahlias, a firework of a flower, reds and oranges shooting outwards.

A cerise poppy, a new edition this year. And welcome too.

Queen Anne's lace. (After the film *The Favourite* we have a whole other vision of Queen Anne now.)

Nigella, white.

Blue lobelia.

Globe alliums, another exploding firework of a flower, the bare straightness of the stem and the perfectly balanced ball of tiny purple flowers both solid and airy has a kind of mathematical grace. You can just look at it.

Bachelor's button – the sweetest of blues. Truly remarkable.

Sweet William.

Lavender.

White violas.

And, either side of the walkway we opened up at the back of the glasshouse into the Wedding Field, where Deirdre got married, this year's addition: two dogwood trees. White star-shaped flowers.

These of course are just what is in bloom right now. There are many more coming, and others already gone; if

I catalogued them all I could fill many pages, but the truth is we spend so much time outside these days that there is little energy left for writing down. The garden grows and changes quicker than I can catch it. But today I wanted to do just that. To catch the *now* of it. To say here, this evening, we are both still alive living among the astonishment of exactly all these flowers.

*

Also, this: by a dispersal magic that all gardeners know, in the polytunnel, among a straight row of lettuces, a number of poppies have shot up. These are neither the small orange ones that self-seed in the paths, nor the giant-cup red and pink ones that occupy the top borders in May. These are opium poppies with strong pale green, near-grey, glaucous stems and exquisite flowers of a kind of magenta mauve. Two things come to me when I see them: first, that in another tunnel they might have been weeded out early on, for they break the lines of the lettuce. They are, in one interpretation, *in the wrong place* – as sometimes in the thirty-four years here Chris and I have thought we are – but they are blooming. The second, that Chris has certainly noticed them from the first moment the seedlings began to appear, and that she has not only allowed them, she has watered and minded them, because she is a person whose life is bound up, helplessly, naturally, with what grows, and it would never occur to her to pinch out anything that might turn into a flower; what would occur to her is in fact the opposite: that these, like all that is growing inside our hedge line, were come into her care, and it was her responsibility to ensure they thrived. It's a small earth philosophy, although Chris wouldn't consider it in such high-flown terms, for her it's not something you have to think about, which is how it's supposed to be, I guess. Still, standing in the polytunnel today and looking at the nine poppies, it seems a kind of blessing to me.

*

I have been spending some portion of the morning thinking about the garden as a living physical expression of the idea of the sacred, and maybe as close as we can get to that on earth. This is an ancient perspective, and in many religions, but is perhaps not an idea for these times, and I have been cautiously trying to figure out just what it is that I believe. It's a complex thought with many side roads, and I have lost my way. But something of that travelling is in my head when I look up this evening and see the light silvering the eastern side of the weeping pear tree.

*

There's a line in one of the psalms, I think, that asks why must the wicked prosper, and it's somewhere in a Hopkins sonnet, or perhaps it's John Donne. In any case, that was the line I was thinking of today when I went down to the lower side of the Grove near the road and saw what had become a mire of nettles. They were chest-high and deepest green, some of the healthiest-looking plants in the whole place. Why, O Lord, why? Of all the plants that might have self-seeded here, why nettles?

In a spirit of penetrating that mystery, I suppose, I decided to reverse my thinking and look upon the nettles as a bounty. So, with gloves and shears, I waded in among them and began the business of harvesting. I cut 100, 200, and brought them up to an empty bin, pushing them in, realising to fill the bin I would need more – and look! We have them. Cutting

300, 400, until the bin was crammed and the broken leaves and stems released a sharp scent. On top of them, I poured buckets of rainwater from the butt until the concoction reached the brim.

Now, there was an undeniable tang of virtue in all this – not to say smugness – saved rainwater and put-to-use nettles to make a rich plant feed. But what of it? Sometimes gardeners have to let themselves small moments if not of actual victory at least of un-defeat.

In a week, we will be able to draw off the first of the feed, in three it will be at full strength, and we will serve the free fertiliser to all that grows in the privileged higher ground that looks down, certainly smugly, at the lowly mire of nettles.

Well, I thought, putting the lid on the bin afterwards, Mr Hopkins or Mr Donne, your question is not so simple.

*

There is this too. One year a particular plant can do extraordinarily well for no known reason. It surprises, and reminds you that for all your care and thought, for all your *intention*, there are things happening that you are only witness to. This year, for instance, the purple lupins outside the conservatory where I write. In the fairly dry May – 'fairly' being a long negotiated but now agreed term between the Creator of Weather and me – they escaped the slugs who come up the garden in the wet and adore them with an adoration that leaves the plants devasted. Instead they matured full and lush and green and sent up their purple spires perfectly straight. Now, they are more or less flawless. They are the TV version of what a lupin should look like. Besides their appearance, their scent! Their scent is a perfume out of the old world. It knocks you back into some other time. I have always been powerfully affected by the sense of smell. It is the hardest thing to capture in writing for it moves so quickly from the physical to the emotional, from the exterior to the interior, bringing memories, stirrings, things too insubstantial to be called memories, a disturbance of veils in the mind. But in any case, if you see a lupin bend to it and breathe it in, is my advice. This scent is staggeringly beautiful, is what strikes me this year as the purple lupin comes into extravagant bloom. I count twenty-eight spires, and for a week, two, they are so perfect I keep thinking: *How did I not notice them last year?* They were there, but only this year do they strike me as truly remarkable.

There is something profoundly consoling in this. Something of: *Your time will come* that speaks to all lives lived in the margins.

The lupins are not this year's only surprise. That purple poppy has self-seeded – I love that coinage, *self-seeded* – not only in the tunnel but in the three raised beds. It has something of against-all-odds in it, something if not quite indomitable, at least of no surrender – in several unlikely places. Chris says she planted one several years ago. It came, and then disappeared. But after the heatwave of last summer and the mild winter, it somehow reseeded in various spots, none of which are where a gardener might want it – in the raised bed of the scallions for instance, even, and most unlikely of all, inside the polytunnel, having travelled in in a barrow of soil or horse dung, or on a wing of a bird high on opium seeds.

These surprises are small reminders that you are part of a living, changing space, and also, that you are not in control of everything, a good thing for me to remember. Another, that you can't imagine what will lift your heart tomorrow. But somewhere, today, it is already starting to happen.

This ground of my ancestors had potatoes and cabbages and onions and carrots and rhubarb growing for years. Every year the same thing. Don't forget the turnips, N says. Right. Yes, they probably grew turnips too, and definitely swedes and maybe some parsnips. Does the soil have a memory? Does it mind now that the front garden is made entirely of lawn and shrubs and roses and grasses and perennials with not a single vegetable? Just all these flowers? I remember my grandfather, PopPop, Joseph A. Breen, who was born in this house and lived here until he went as a young man to America, saying he loved rhubarb. Could it be from his childhood here? Rhubarb would have been in most Irish country gardens then. It's virtually pest free and decorative and low maintenance. (Aspects we increasingly love in our garden.) I remember Mary Breen showing me how to take a stem from the plant. How you slide your thumb and finger down both sides of the stem to as near the base as possible and gently break it away. No cutting. And how she told me the plant needed to be deeply manured. Manure seems to be a cure for everything. (Well not if you're into vegan gardening.) But way back in the early days we didn't want two giant rhubarb patches outside our front door with their heavy scent of cow dung. So we dispatched them to elsewhere. Now it's time to return rhubarb to the garden and we have planted one near the blueberry bushes where the soil is moisture retentive (duh – it's the boggy west of Ireland) and we've added half a bag of well-rotted manure. Until next spring then…

Midsummer. A wind from the east that the plants are not expecting – the petals fly. A peony bloom of absolute perfection yesterday this morning bows to kiss the ground. The sky a pale grey moving like an opaque sea overhead, the treetops dancing with the weight of the leaves. *Where O where is the sun, Saint John?*

In the lungful gusting and sudden bursts of clatter rain, Chris and I are outside again with more bamboo canes and green twine. Up to today, I now realise, there has been so much perfection, plants flowered and stood in a kind of formal grace. Evenings you could pause at the front door and see the whole of the garden in a standing stillness with scatters of colour like brushstrokes, or dancers, birds darting against the late light, and everything held as in a composition over thirty years in the making. And under the spell of that, you forget how frail and vulnerable the whole thing is.

Which is what this morning shows, as we go from plant to plant, straightening, tying up further along the quadrant of bamboos. 'What are we supposed to do?' Chris asks, 'Just leave them to be knocked over?'

I have learned the wisdom of saying nothing at such moments. I myself dislike the sight of the canes sticking up above the extraordinary blooms of the poppies, the artifice of supporting nature, but what if nature will otherwise be crushed, the peony flowers lie on the ground? An explosion in the centre of the poppies. Somewhere between the nature of the weather here and human nature, the impulse to support, foster, aid, is in miniature every gardener's dilemma, and finding the balance between nature and the 'natural' is part of each day's learning.

'You learn from your soil the place you are gardening,' a friend had said. 'You learn what you can grow, and what you can't.' And certainly, that's true. But there's something else too. Chris is always, always, trying to improve the soil here, and although that, I now realise, will be a lifelong endeavour, she has done so and continues to, so that some things that once didn't grow well now do. So again, what is nature and what natural are questions not so easily resolved.

By mid-morning we have everything staked. The rain batters down and the wind hurly-burlys everything back and forth. Only tomorrow, when the midsummer storm passes, will we see what's standing.

zinnia bud

7

July

'Has the cuckoo gone already?'

Always a little fall of the heart each year in the fact of that. Summer is moving on.

But this morning, darting out of the open doorway of the cabin, one, two, three, four, and yes five, young swallows. A second hatching, or from a nest we hadn't seen, they swoop out and curve around and alight in the branches of the dead cherry. It flickers alive with them for moments. Then they are off again. Back around the garden and in through the doorway once more. My delight is in theirs, which is so apparent it's palpable.

The hori hori knife is the best tool, ever! If you can only have one tool in the garden, go for a hori hori. It's from Japan and apparently the sound 'hori-hori' is the onomatopoeia for a digging sound. (Well, not to my ear, but in Japanese maybe. It means 'to dig'.) One side is serrated and cuts like a small hand saw. It's perfect for long dandelion roots. It transplants, cuts grass away from flagstones. Splits perennials. And it has a strong sturdy wooden handle. How have I gardened so long without it?

━❧━

There is not a small challenge in trying to be free. That sounds grander than I intend, but there's truth in it too. It seems to me that what we have been trying to do here for thirty-four years now is to invent a life. I mean invent a life that we want to live, and to do so while figuring out what exactly that is. That there is no rulebook or guidelines for this kind of living, was something we quickly became aware of.

(What are you supposed to do first thing on a Monday morning? To which the true answer here is: *Is it Monday?)*

Now I want to be careful here not to write with a glory-pen. There's a deep security in living inside the norms of job, workplace, work hours and all the rest, and I was glad of that security years ago when the children were small and I was doing some teaching. But I was always aware of a confinement in the classroom and the curriculum, and that something had been traded in exchange for the constancy of the pay cheque. I don't mean to denigrate that here. The teaching was a privilege and often a great pleasure; I loved trying to communicate my own love for literature, but there was something in me that felt compromised, that in some way I was not *supposed* to stay doing it. Sounds odd, I know. A prompting in the blood or the brain that may be nothing at all. In any case, when I stepped away from the teaching it was not because we had any security or savings that would support us, it was because I felt we hadn't come here to Kiltumper for that. As I've said, we had come largely on the purest Romantic impulse – I use the capital R advisedly, meaning not as the word is used for lovers, although we were also that, but as idealists, dreamers, believers in something higher, better, than the life we were living in New York. And so, it was time to return to the pursuit of that.

And pursuing is a good way of saying it. In one way, that life is something that is always just ahead.

This, I learned, is an integral part of gardening too, this sense of how next year's garden will always be an improvement, you'll have learned that little bit more about the soil, the conditions, the plants, or all three, and the wisdom of what it is to be just living in a place will inform what next year's garden looks like. As I've noted in January, that it'll be better next year, is the winter dream of all gardeners. (*Maybe the winter dream of all plants and trees and living things, too.*) In any case, there is a not insignificant challenge to maintain this perspective when reality meets that dreaming, both in gardening and in living.

I'm not being clear enough here.

What I'm getting at is it's one thing to dream a life, another to try and live it.

So, while in Kiltumper we are free to invent what we do each day, free to go to walk the beach at Doughmore if the day is fine, to spend the whole day in the garden, or read a book, we are also living within the constant uncertainty that there will be any money to go on living like this next year. That's the reality. It would be the same for anyone else. And sometimes you can fall into a darkness in which you berate yourself for even thinking you should be trying to live this way. What gives you the right? is a common enough accusation at such moments; so too the thought that it is a vanity to separate yourself out from everyone else who is heading to the office, the school or the factory. And at such moments there are no easy answers, and you work, whether at writing or in the garden, or helping Chris cope with pain, with that sense of spiritual disquiet and profound aloneness that can only be assuaged by focusing on the concrete.

Well, Chris and I have been juggling these thoughts for decades now.

What is required is a level of faith. First, you have to believe that a life like this is worth living. That it is what you should be doing. I mean 'you' in the narrowest sense, I'm not being prescriptive, not for one instant advocating that this is what anyone else should be doing. It is too hard for one thing. Not everyone can be this lucky is another. And another, even more rare, that it requires two to have the same shared vision, and, importantly, accept the consequences.

When we arrived in Kiltumper, I knew nothing at all about gardening or land. It was Chris who said we are choosing not just to live in this place, we are choosing a way to live here. And that way of life has evolved over the decades to include the writing and the drawing and the gardening and raising two extraordinary children.

Now, while the writing does provide some income (sporadic, uneven, uncertain, the adjectives that most apply) the garden provides none. And so, the hours we spend in it each day have to be computed differently, where the value is measured on an inner scale, and not by the question of how much did we make from all that work this afternoon?

I don't mean to suggest we are always able to see this with perfect equanimity. We are not saints. Often enough, we fall down, lose hope, and are defeated by the many difficulties of sustaining this life, by its uncertainty, fragility, or the overwhelming loneliness of a garden in November rain. In my notebook one day this month, a single line has in capsule form a whole volume: *Being free is life's most difficult path.*

And well, there's nothing new about this. In fact, there's something very old. The point is that in our case it's a choice, we were free and privileged to be able to choose it, and I like to believe that from the fact of the choosing flows something that sustains us through the difficult and uncertain times.

There is also the physical, sensual reality of the place itself. Because we have moved from the idea of a garden as a hobby to one where it is an essential component of our living, we have the sense of being part of it as it grows all around us. And the sustaining thing is its beauty. Just that. And that's a lot. The real value of living with the beautiful is something I have thought about quite a bit. What it means to be in the company of the beautiful as a kind of soul-medicine is something John O'Donohue wrote about, both for the sensual delight and the spirit-lift, the response that begins in the earth and travels instantly beyond. And well, in that light, the garden is the most valuable thing we have in our life here.

～✎

I took a walk this evening while N is away. The construction of the turbine bases is finished for today. I followed the small road that goes between the Big Meadow behind our house and the Bog Meadow to the east. I climbed over the gate into 'Lower Tumper'. It belongs to my father but, before he decided to plant the fields with hardwood trees, it was home to our grazing cows and their calves. Years ago that is, back in the days N and I tried our hand at farming, when we first arrived. Then, we'd walk up these hill fields, Upper and Lower Tumper, to count our small herd and stand a moment and watch them and listen to the quiet sounds the land itself seemed to make.

I walked across the thistle and tall grass into the heart of the hardwood forest when some movement at the edge of my vision made me stop. I turned in time to see a fox climbing over a stone wall and down into the field at the back of our house. Two brown mares belonging to our friend M were grazing. The fox stopped still and slowly turned his head to me. Both of us then seemingly seduced by the stillness and beauty of the warm evening. In a snapshot second it was pastoral picture perfect. Alone on the hill I recalled Frost's poem, 'Stopping by Woods on a Snowy Evening'. Here I was, a single and yet not-so-single viewer in the silent landscape of ash and oak, of fox and horse, of meadow and sky, and I felt gratitude.

And yet and yet and yet…

What will it sound like here when the turbines are chopping the air with their forty-metre blades only 200 metres from where the fox and I are standing now, perfectly still?

***

Most mid-mornings now, while I am writing in the conservatory, I see Chris pass with secateurs and colander. It is one thing, in the first heady rush of spring, to set seeds and put new small plants in the ground, another to see them coming through May, to take the pleasure that comes from participation in the renewal of the world once more, but it is another still to do the harvesting. This year Chris, who has a self-critic never far away, is determined not to let lettuces, spinach and the rest bolt because she was reluctant to cut them while they were not yet at their peak.

This year a new policy, cut as they grow. Take outer leaves. And keep taking them.

There's another small act of faith involved in this, the plant will survive, and more, keep on producing.

So, these days, there is something fresh from the garden in all meals. There is salad every lunchtime.

Today, as Chris passes back across the window with the salad leaves harvested for lunch, her look in to me says, 'See?'

I have just been writing about trying to maintain the balance of belief in this life, but when I look up and see her standing there with the small harvest I do see.

'They're not great,' she says, 'but still. Right?'

It takes me a moment to come from the abstract into the real, from all this thinking and writing into what is actual, there in front of me, this life.

'Right.'

'I'll get it ready for lunch,' she says.

'I'll be there in three minutes.'

I watch her go across the front of the house with the salad leaves and calendula flowers and blue petals of borage and the tails of three scallions at various angles, and the simplicity and grace of that, the *goodness*, I want to say, defeats all attempts to write about meaning or the idea of freedoms and choices of how to live.

Remember this moment, I write in my notebook.

A moment like this may be the only wealth of our life here, but it is a real one. I close the notebook and go to help with the salad.

*

The fourth of July. The hammering rock-breaking up on the site for the turbines wakes us at 7:07 a.m. It will be going now until 6:30 this evening. So, although the weather is beautiful and you want to be out in the garden all day there is a real challenge in how to do that and ignore what is in the air and in your ears. I have devised stratagems for diversion: in the house turn the radio or the music on first thing and keep it on all day; set up a speaker for Chris out in the glasshouse.

In the pause between songs, as I sit here, with all windows and doors closed, I can hear the constant *clack-clack-clack* of a rock-breaker, punctuated by a harsh scratching and then a squealing sound as rocks are shattered, dragged, pounded.

The turbine site they are working on is the one nearest us, on the slope of the hill of Tumper (locally pronounced *Tomb-per* or *Chew-mm-per*). Kiltumper is an unusual place name hereabouts, and when we first arrived here, I did some looking into it.

'Chew-mm-per was a giant, Niall,' Mary Breen told us. 'He's buried above on the top of the hill. It's written up in a book.'

Mary wasn't wrong. That book was an archaeological survey of the tombs and monuments of Ireland, carried out in the early nineteenth century. And sure enough, in there is listed the townland of Kiltumper, whose only notable archaeological feature is Tumper's Grave, an east–west mound on the crest of one of the hill fields of our farm, with Tumper's Stone, a large standing stone at the head of it. That stone, sadly, was gone by the time we arrived here – it was taken for some farm building somewhere or is a pier for a gate – but the mound remains. It's about twenty feet across and thirty long, so not a giant giant. (Then again, I think I remember hearing that people in the iron age were shorter, so maybe anything over six feet was giant.) Of course, the archaeological book doesn't credit him as a giant, but a chieftain, which is fair enough. Now, to anyone whose mind tilts towards story, as mine does, it wasn't a far leap to connect Tumper-the-giant, Tumper-the-chieftain, with the Long Breens, and to feel some kinship with the tall lad up on the hill. It felt fitting somehow. Chris being related to royalty also doesn't escape me.

So now, today, across the boundary wall from the field with the tomb, as the machines encounter the rock face underneath the grass of the hill of Tumper, and find it resistant to the digging out of a foundation for the turbine, the cacophony of clacking and scratching and banging strikes me as eloquent. To someone of my peculiar imagination, there's something of a plaintive earth-song in it, there's objection and resistance. There's also the sense of Tumper-the-giant's outrage at having the face of his hill dug away.

The old lad is not surrendering easily.

The banging continues all day.

I turn Jackson Browne up to ten.

*

The many pleasures of a tree. In this month the judge's wig of the weeping pear, silvered leaves piled on top of each other, lends it an august air as it looks back at me this morning through the cathedral arches of the conservatory windows. It doesn't fruit, fruit is a frivolity, the judge stands upright and still and pale and aloof.

But for all its august-ness, walk down to it, put your palm onto its leaves, and discover: it is a softie.

Summer advances and the cuckoo's call has changed. It has a sort of guttural but lonesome break in the middle of its cu-ckoo. The rhyme goes, 'The cuckoo comes in April/ He sings his song in May/ In the middle of June he changes his tune/ And in July he flies away.' What we hear when 'he sings his song in May' – the nearly absurdly literal cu-ckoo, cu-ckoo – is the male calling for a mate. Once he's found her he stops calling. (That figures.) The 'change' we hear in the middle of June turns out not to be the male cuckoo changing his tune – he's gone and flown back to Africa – but the female cuckoo calling. Perhaps she's lonesome for him.

We love the call of the cuckoo even when we know the female has laid her egg in some other bird's nest and is too lazy to raise her offspring and worse has displaced the eggs that belonged there. You kind of start to dislike her, but you forgive her because the call of the cuckoo is magic, isn't it?

⌒∿

'Well, it's my favourite thing to do. Clear a space. I did it in the woods growing up. Could do it for hours. I'd leave the house, escape what was going on there, go out into the trees and find some place and I'd start clearing. But what was I thinking today?' Chris lies on the pebbles, exhausted after hours of working. Beside her on the ground, her secateurs. Behind and to the right, a very large mound of weeds and branches, and behind them a cleared patch of earth.

The cleared space, I am thinking, is one of those elemental maybe original instincts in mankind. We never think of Eden as a garden that needed clearing or maintenance; I suppose the logic is it was Before the Fall and therefore perfect, but if it was a *living* place, if the plants in it grew, then they would grow too large and crowd each other and need pruning and dividing and all the rest of it. If they didn't grow, it wasn't a garden as we know it. Well, where my thinking on this takes me is: a garden, as I understand it, is a thing that is inextricable both from change and from human activity; it needs us to exist. A wilderness is not a garden. Even the apparently paradoxical 'wilderness garden' is an act of creation and the result of thought. (Out behind the garden proper, in the Wedding Field, we have decided to let the grass and wild flowers grow, and only mow a path through and around it for walking. So now in July we do have 'a wilderness' that grows up to Chris's waist and she likes to walk through it in the evening before finally leaving the garden and coming in for the night. But the mown path is what makes it a 'garden', if it can be considered one. The path is what invites you. And part of that invitation is that when you see the opening through the tall grasses you connect to the mind that first imagined it and think: *Look, here's a way to go.* In other words, it's a human connection.)

So, clearing a space, either for new planting or to lessen crowding, is an integral part of what people do when they pick up trowel or fork. Very often, it is what Chris is doing when I look up from writing in the morning and her head appears in the lower bed or over by cabins, or when I bring her a mug of tea and call her name and the answer comes from the ditch across the road.

'What exactly are you doing?'

'Just clearing. Can you tell? You probably can't tell.'

'Oh, I can.'

'Really?'

'Definitely. Right there.'

'And there, and there, and all along there. Well, the point being, this soil seems to prefer weeds to anything else and *someone* has to dig them up.'

'Yes. I can see.'

'Only because I pointed it out.'

'No, looks much … *clear-er*.'

And it does. You can't quite get to the bottom of why. At least I can't. It's more than aesthetics, more than inherited ideas, the imposition of order or the need to control. But somewhere in there is that idea of the inextricable nature of mankind and gardens, the silent covenant, whereby the gardener clears a space *for plants*, and invariably, through enhanced growth, those plants repay the gardener for the work done on their behalf. In all aspects of our lives a cleared space always brings some mental ease, but when it is done for plants and flowers it seems to have something indefinably but intrinsically *good* in it, to me at least. And looking at Chris, lying exhausted on the pebbles today after hours of clearing, that's all I can think to say:

'Looks good.'

Delphinium comes from the Greek delphis, *which means dolphin, because the closed buds look like a dolphin's nose. The blue of the Kiltumper dolphin is a cerulean blue, one of the Pacific Giant Blues, and it grows nearly five feet tall. ('Blue skies smiling at me, nothing but blue skies do I see.')*

As Carol Klein writes, 'a true, rich blue is rare in the garden.' Her favourite blue flower is Anchusa azurea *'Loddon Royalist'. As with so many plants that are prized, they're difficult to grow, especially in heavy clay soil. The best way to propagate anchusa is from root cuttings because even the seeds don't ring true. At last year's Chelsea Flower Show in the Great Pavilion, I got some anchusa seeds and nursed them during the winter. Of the three that seeded, one failed. Of the two I planted out, only one bloomed. Carol says on her clay soil she gets two years out of an anchusa. Next May I'll either be chuffed or disappointed.*

Mid-July growth overtakes all gardeners. There is just too much happening, and you have to adjust to the idea that the order of April and early May is overwhelmed now. Plants are almost growing as you look at them. The weather has been warmer and drier than usual – although we are aware the usual may be changing – and seems to have provoked a flurry of blossom. Everywhere the garden is crowded with colour and flower. Each window of the house has a view with something colourful to look at.

What is it like to be living *inside* all this growth and life and colour for over thirty years? How surprising it is that we haven't ever become used to it to the point where we no longer see it as extraordinary. It is, and we do. There is hardly a day when one or the other of us doesn't comment on some plant or flower whose character or quality has caught our attention that day. I know this sounds an exaggeration – and it is true I exaggerate, and, as I've probably made clear, to Chris's mind, overly look on the bright side – but actually today I was just struck by the sheer abundance of the place in July, so much so that I tried to remember last July and asked myself if the garden looked this well then? I don't know if it is my age or my character, but whichever, I couldn't get a solid memory of last July. (I wonder if this is true of other gardeners? Does one year's spring and summer garden vanish in memory as the next one comes? Is the garden always just this one, in front of you, right now? It seems so, at least to me, today.)

The truth is, the heady fullness of the garden in July has something incomprehensible about it. It is too sensual to be comprehended.

I am in the quiet rapture of this thought when Chris comes to the door of the conservatory where I am writing. She has the flushed face of summer work.

'The weeds are growing too fast,' she says. She turns and looks at the front bed, in case she catches them growing behind her back. 'I hope you're not just writing that it's wonderful, are you?'

'But it is.'

She smiles her extraordinary smile. 'You!' she says. And we're both near enough to laughing. Something of our whole life together is in this, my dreaming and her truthfulness. The necessary components of all gardens? (And a good marriage perhaps?) My mind will come back to this later, I know; in my notebook I jot down something, but midsummer is too busy now for anything other than notes on the fly. There will be plenty of dark and wet days ahead for rumination and philosophy.

Chris lets the acknowledgement of the wonderful sit for a moment, then she can't help herself. She turns to me. 'For balance, you know, where we made a list of all that was in bloom in June?'

'Yes.'

'Well we should make a list of all the weeds we have growing, too.'

*

Delphinium, I learn today, is in the buttercup family.
Really?
Large family so.
Yup. *Ranunculaceae.* (Over 2,000 known species.)

*

Saint John, I shouldn't have doubted you. Sunshine in Kiltumper. Days and days of warm dry weather and the place is transformed. A heatwave burns in Europe; out here at the fringe we are spared the brutal intensity, but get a magnificent settled period of high pressure. There is not a breath of wind. Each plant and flower stands so still they could be painted versions of their best selves. The house has its doors and windows open as if to embrace the invisible. Because the sun doesn't go down until very late in the

146

evening there is a sense of living in light. Your senses are aware of it. Because here our bodies are accustomed to only the one or two glorious days together – three is a heatwave, and five a good summer (I exaggerate, but there's truth in it) – there's an adjustment you have to make when the sun is there every day. That adjustment is mostly about relaxing into it, accepting the gift, something I am poor at, but trying to do better.

'Feels like actual summer,' Chris says. In the garden, she wears a straw hat that looks Provençal.

Summer, I think, sounding it in my mind early the following morning when I take a mug of tea outside into the garden where the stillness of all the blooms is contrasted with hectic bee sounds. Summer. The deep soft 'u' kissing the consonance of the two 'm's and the purr of that 'r'. *It is a sound exactly as rich as the taste of a strawberry*, is what I think to write in my notebook, but almost immediately think the image inadequate, as all images are, and decide that, by language-magic, summer can only be captured by the sound of the word *summer* itself in your mouth when the sun is full on your face. There's nothing for it but to move out down the garden and raise my face to the morning sun.

'Summer. Summer.'

And for sunlit moments then, I am a man in his sixties inside his garden talking to the air.

*

Of course, with the heat, and after last year, we are watching the well, sparing our dishwater which we carry out in basins in the evenings to whichever pots Chris adjudicates most need it. The two water butts, each brimming with rainwater, are invaluable, and as you fill and carry the watering cans you can't help but both castigate yourself for all the rain you haven't caught over the winter and congratulate yourself at the same time. Which may be a fair encapsulation of human nature.

The back meadow is high and full and turning blonder by the moment. In the belief that the heatwave will hold a few more days, our great friend Martin comes to make hay. Once the grass is down, there's the summer clockwork of him circling the meadow in his tractor with the hay-turner. It clicks on for a time and then he goes away, comes back in the evening to turn it again. Hay is a hopeful kind of thing; it is captured sunlight really, the smell of it sweet and deep and ancient. I stand at the gate from the Wedding Field into the Big Meadow and just look. The mowing and turning has left gathered rows of hay waiting for baling. The rays, as they call them here – I'm not sure about elsewhere, it's a lovely word for them, containing both sunlight and geometry – are a kind of clock too. If you were looking from God's perspective, you'd see the meadow marked with concentric circles of gold (lopsided ones in the case of our meadow, there's some boggy ground in that top corner which doesn't get cut, but still) and they'd speak both to the beneficence of nature and the climax of another summer. *That year the hay was good,* a sentence full of an implicit gratitude that I can picture as that antique thing, like the hay ropes they used to make, only this one winding from our back meadow up into the sky.

When we first arrived in Clare, it was a threshold-time. Old farming practices still endured alongside the more mechanised ones that were coming. At that time, in this townland, many farmers still made hay. Silage-making was coming; it would be widespread within a few years as one bad summer followed the next, but in that first July it was all hay that was saved. Weather forecasting wasn't what it is now, and for the three dry days needed to save the hay, farmers watched the sky and spoke to each other and gambled on when to cut. Then, once one of them decided to 'go-for-hay' the neighbouring farmers were on hand to help. There was a communal quality that was completely foreign to us. I can still remember the evening Joeso stopped by, having collected two buckets of drinking water from the well, and told us Michael Downes was making hay tomorrow. He didn't say the rest of it. He didn't say: 'And he could do with your help.' But I understood. And the following day I walked down the road as if for an appointment – wearing wellingtons that were not what was required for this hot job – and was welcomed warmly at Michael's haybarn by several of his sons none of whom were yet in their teens and who had some fun instructing me in the use of the two-pronged hay fork. After that, what I remember is the busyness of it, the engine of summer as Michael's and the neighbours' tractors came one after the other into the yard carrying high stacks of hay, backing in and landing them more or less beside the barn for me and Joeso to fork up and into the barn, the young boys scrambling about above us in the golden playground with the job of making sure there were no air-gaps and the hay was packed tight.

'You're the right height for this job,' Michael said, with a grin, watching my earnest attempts to fork the hay as high as possible – not seeing the one time the same earnestness resulted in me throwing the fork as well high into the barn, but not killing any of his sons.

Well, it is dusty and warm and golden work, and when the barn is packed full to the roof with hay your arms aching, your eyelids burning, your throat a tinderbox, the sight of the harvest as you look up at it is simply marvellous. I've never forgotten it. That hay is saved in memory.

And I was thinking of that memory this evening after Martin's tractor had gone and I was standing at the gate looking at the fallen meadow, thinking too of something I had read in Wendell Berry: 'In a viable neighbourhood, neighbours ask themselves what they can do to provide for one another, and they find answers that they *and their place* can afford. This, and nothing else, is the practice of neighbourhood.' And certainly, in that remembered time, I had known exactly that, and maybe only appreciated now the gift of it. That's all right too. For the most part, the farming is more mechanised, organised, contracted now, but there are still many living examples of the same thing here. Knowing he had the hay-turner attached to his tractor, Martin had a call from Michael to know could he turn his hay and that's where he had gone now.

This evening in the Big Meadow, between the rays of hay, the ground was pale, the air trafficked with golden motes that flew and floated in the late sunlight, a kind of making visible, it seemed to me. I can't quite describe it, but it was to do with a sense of something as old as our existence on the planet, this moment of complicity between air and land, between sunlight and soil, *hay*, just that.

Just the goodness of that.

*

Tonight, a feast of vegetables for dinner all from the garden: courgette, yellow squash, kale, spinach, basil, shallots, parsley and the first of the four different kinds of peas from the mass of them now on Pythagoras's pea fence.

Good lad, Pythagoras.

*

Just now, nine bees going in and out of the coming-to-bloom purple spires of an astilbe (ah, still-bees, yes) not four feet from where I am writing this.

We are living in a crisis time for bees now. Seems extraordinary to think of. The depletion of the bee population is a man-made phenomenon, and, indisputably, one of the most damning indictments of how we treat the planet. To be clear, I am not a scientist and these pages are not a place to find the figures and the proofs, but the case that first seemed alarmist is incontrovertible at this stage, and so too the essential role of the bees in our continuing life on earth. Such large statements can make a person glaze over and throw up their hands. The problem so immense, the situation already so far gone, that it's easy to fall into a resigned position of *What-to-do?* and just hope for the best.

I'll admit that when I first heard of the predicament of the bees, I didn't pay it much attention. I can't say I was a fan of bees, as such, and ignorance meant I didn't lose any sleep over it. When I first read the line in Yeats about 'the bee-loud glade' in his dream-island of Innisfree I was probably sixteen, and in suburban Dublin had no lived sense of what that was, or the resonances, literal and metaphoric. It was only when I came to live in Kiltumper in my late twenties that I heard the bees in the Grove and was startled both by the actual bee-loudness and the accuracy of the poet's image.

On a personal scale, the answer to 'what-to-do?' is simple: flowers. First, help the growth of wild flowers in meadows and wild grass. To this end, as already noted, we left the half-acre of the Wedding Field uncut but for a narrow path around the edge for walking, the rest left to itself. In the autumn we will begin the process of trying to encourage and introduce more wild flowers in it. This is not as easy as it sounds, but Chris is determined.

Besides this, of course, there are the very many flowers already in the garden. And this year I am noticing the bees in amongst them more. It is not that there are more bees, just that I have more awareness.

So, just now, watching the bees at the astilbe four feet away through the glass, I am reminded of all this. The bigger picture, and the smaller one. Their pattern is no pattern but the chaotic one of the impulses of pleasure. A bee leaves one spire, goes to not the next but the third one over, spends only a half second, goes to one four over, stays there for a count of eight as I watch, comes back to the second one it missed before. Watching the bee-delirium of them, you can nearly suck the sweetness yourself. They are drunk on it in the middle of the morning.

They live here with us too, is what I note. *They know this is a garden for them.*

Now that the diggers have clawed away the face of the hill in the next field over from our Big Meadow, and what was green is now grey shale and gravel and giant orange cranes, we can see the exact siting of where the turbine nearest to us will be. They will be full on, full frontal, in our faces.

A very kind nurseryman in Bunratty gave us a young hornbeam – a beautiful tree which doesn't mind wet earth too much. N and I gauge the sightline from the not-yet-planted turbine to the kitchen window and decide on a place to plant the nurseryman's gift. Eventually, long after we are gone, the hornbeam will help screen the sight of the spinning blades catching your eye every time you exit the car and spot it from the kitchen window as you look north-west.

Part of me believes I am planting this for my children. Another part of me believes they will never live here. Not now. Not with those noisy things towering above our house, our five-acre field, my father's forestry, the bog, my garden.

Now, there's not a small irony in planting a carbon-eating tree to try and block off the 'green energy' of a turbine, nor in planting a tree when many old ones along the road will have to be felled to make space for the one day that blades are brought in. But we do not speak of that while we work, clearing the old grass back, digging out the planting hole, hammering in a support stake, filling the hole with water, spreading the roots of the hornbeam when Chris takes it from the pot, laying the tree in along the stake and filling a not-too-rich compost in around it, hand-pressing the earth in snug, making sure everything is just so, before stepping back to look at it, the newest tree in Kiltumper.

The next lorry full of gravel for the site barrels along the road outside. It makes Chris's stomach churn each time.

*

A delivery man stops at the front gate enquiring of us the whereabouts of one of our neighbours. In a South African accent, he comments on how busy our road is with lorries and we tell him it is for a wind farm.

'The turbines are everywhere, eh?' he says. 'I don't like the things. They destroy the green, don't they? We lived in Rosscarbery in Cork and they put one in a kilometre and a half away and all you could hear when you were outside was a constant whoop-whoop-whoop, like that, you know? You couldn't escape it.'

He doesn't know we will have one 500 metres away. He doesn't know that when he drives away Chris will be upset, that we will go into the house and be able to say nothing to each other for a bit, and that what he has said will bring us face-to-face with the thought we don't want to consider: *Will we have to leave here? Will everything we have done over the past thirty-four years to make this place and make our life here, be destroyed?*

To escape the feeling, we drive to the ocean at Doughmore. There is not a single other person on that magnificent beach. All arguments exhausted, we walk in silence. If we have to leave, if we have to give up our life here, I think, and then I have to let the rest of that thought end in ellipsis. We walk on with the sea breaking itself on the shore.

8

August

That the climate is changing is a fact that anyone who works with soil and plants knows. In years gone by, I have often been surprised during the brief periods when I have done press for a new novel by journalists asking me if it really rains in Clare as much as it does in my books. The question was genuine, and in time I realised that it was being asked by people who largely lived and worked indoors, who in the course of a normal day spent few hours actually outside. Weather then was a factor confined to the morning commute, lunchtimes, weekends, and in realising this a couple of things struck me, first that I was fortunate enough to know the rain at first hand, and second that for most people climate change would remain an abstraction until the streets were flooded or the taps yielded no water. This is a terrible thought. And I reproach myself for it by saying I sound like a grumpy old man despairing of the future. I am older than I was, I concede, but not grumpy. Saddened though, yes.

It occurs to me that when cataclysms have happened in the past, the world wars, Chernobyl, say, there has been a sense that they have been *over there*, elsewhere, another time. The twenty-four-hour news we live with now brings us first-person reports of catastrophes daily, and yet, largely because they are on television, they take on only the reality of television shows, which is to say, they are unreal. For the past decade or so we have been hearing of climate change, like a gathering storm, hearing that the glaciers are melting, or the rainforests are being plundered or on fire. And while there can be few who haven't heard this, there has also been a sense of it occurring in that elsewhere place. Which, in some ways, is understandable. We are not in the Arctic or Brazil or California or Australia… I say this partly to excuse my own lack of urgency.

Today, the latest United Nations report on climate change was published, and reading it here in the green of Kiltumper, the feeling I have is of self-recrimination, and then loss. A sickening feeling of loss. The world is losing. The deforestation of the Amazon rainforest continues, the last glaciers in Iceland are starting to melt, the oceans are flecked with microplastics. The report has many pages of defeats on all fronts, and the argument is overwhelming.

In trying to come to terms then with the failure of all governments, all lawmakers, all who have been in positions to address this, it is hard not to succumb to the dejected position that unless there is profit in it, nothing is going to be done. The problem can be acknowledged, but it is being passed along to the next generation to solve. By the time the term 'climate change' has become 'climate catastrophe', it will be too late.

One of the phrases in the report that struck me most was 'human-induced degradation of land'. When I read that, I had to stop and think about it. I often come to a deeper sense of a thing through language, and that phrase was deeply disturbing. Not that I didn't know that human beings had always exploited the ground they lived on, but *degradation of land* seemed appalling. At the heart of it is an absence of *love* of the world. You do not degrade what you love.

If I was to pick one thing that I have learned from our years here, it would be love of the world, in particular, of this place inside our hedge line. In our every day, Chris and I are trying to achieve the opposite of degradation, and in this I haven't thought we are any different to the most of mankind. I don't award us any medals, for it is natural, and what is natural feels good. This is the premise anyway. It underlies all human endeavour.

I do believe that in a hundred years, the future, if there is one, will judge our more-than-carelessness with a cold eye. We will be considered shameful, *because we knew*, each report after the next, and because we had the chance to stop it.

Over thirty years ago now, Chris and I published the first of four books about our life in County Clare. We wanted to call it *In Kiltumper*, but the New York publisher said nobody would know where that was. *O Come Ye Back*, as it became, recounted our first year, in the four rooms of the house where Chris's grandfather had been born. It was a true, accurate account, but also an innocent one, as was befitting a book written by young people dreaming themselves into the world. And speaking to that, running through that book as a sort of refrain was the thing Chris often asked of me, first in New York after we had made the decision to leave in a hundred days – any longer and we might have changed our minds – and then during some of the fraught and lonely days in that first year: *Tell me how it's going to be, Niall?* She would ask the question, and though I did not of course know the answer, I knew that an answer was called for, and I told a version of our future, an imagined story that became real for the duration of its telling, and banished, for the time being, the doubts. That version of our life we would then try and live, and in this way be living within the light of a dream, with all the hope, and the failures, of that. Still, there was always this sense of creating a life, around a house and garden, and it occurs to me now, as it did not then, that all of this dreaming and making was built upon the certainty of the world itself continuing. Of things in the big picture staying as they were. We did not at any moment wonder, what if the planet itself is failing? As I said, we were young, and, unlike now, such thoughts were not in the jurisdiction of youth, then.

Tell me how it's going to be, Niall.

I could do as good a job at talking us into a future, if asked today.

But there is hope, I say to Chris. There is hope in the children's movement, the school protests this past year, and the sense that the young will no longer let the old carry on destroying the planet for them. In her small apartment in Brooklyn, our daughter keeps all her compost in a container in the freezer until Saturday when she can bring it to the communal composter that comes to the farmers' market. People are choosing not to buy vegetables that come in plastic wrapping. There is hope in the young, I say. And I do say that believing that, otherwise, this will be the last century. The last century of mankind able to survive on a single planet. The twenty-first will be the last, or the one where humanity finally comes of age, and starts to behave responsibly.

The report leaves me darkly saddened. Not least because I know I am late to this, and could have done more here over the years, to lessen use of plastic, manage fuel better, there are dozens of things, all of which come to love, I think, by which

I mean not taking the world for granted, and generally living with more awareness of its beauty, more love.

Now, it seems to me if there is a common enough response in the face of such reports, it is to be overwhelmed, and with this, to feel a sense of impotency. What good can we do? We are only two people. And, this extends at once to, what does it matter what Ireland does, it is only four and half million people? Look at America, look at China. Your spirit throws up its hands, the earth seems doomed, to be drilled, fracked, fecked, until it can give no more. It takes some work to counter this. And as I've said, in our version here in Kiltumper, it is to think small. Think only of the bit of planet we are in charge of.

I go outside. There is rain streaming down outside, rains that seem heavier to me this year. My sense is the summer has veered between dry hot weather and then built-up cloudbursts. And because I am the way I am, such things seem to be saying something. I know this is subjective and I am too impressionable, and all the rest. Fair enough. But still, when I go out in the rain today after reading the report, that is the feeling, *This rain is too heavy*, too forceful for the first days of August, and I stand and let the rain come down on me.

*

The longer I have been writing, the more earthed I have become in a fiction of place. That geography and history are in fact inseparable, if not actually the same thing.

The idea I consider while working in the garden today is that no human act passes without leaving a trace on the place where it not only happened, but which was an implicit part of it.

This is an idea I could take for a long walk.

*

The defeats of August.

'Wherever humans garden magnificently, there are magnificent heartbreaks.' Henry Mitchell.

In equal measure these days, abundance and dying. The month is balanced between the two, so many plants heavy with bloom and on the point of going over at the same time. I love that gardening term, *going over*, both accurate and kind. There's gentleness, understanding and appreciation in it. The plants do literally go over with the weight of what they produce, and something of this stays with me this morning as Chris and I go about once more like medics, only with twine and canes, tying up what the downpours have done to the dahlia heads. One particularly spectacular mauve dahlia head is larger than a child's face and bows to the ground with water weight. If not lifted, and shaken a little, it will rot away quickly.

'Look at it, just look at it,' Chris says. Her eyes are shining with wonder, as though she has never seen it before, and when you look at it, individually like this, you think you haven't seen it before either. So much of gardening for me is just seeing, and I am aware that in this I am so fortunate to be able to share Chris's eyes. 'There is vision, and there is *vision* too,' my old friend, the late Father Michael Paul Gallagher used to say to me. Well, both are here in this moment with the dahlia.

But, inseparable from its perfection, is the imminence of the dahlia's decline. Tomorrow it will have passed this moment, and begun to go over. As is characteristic, Chris battles against this, and I accept it. She shakes the rain off the half-budded sisters. That this may have something to do with how she has come this far through the cancer is not lost on me; it is a strong impulse, and may be in all gardeners, this desire not to give up on a living thing while there is still living in it. The thought of that moves me as we go from one rain-washed flower to another tending to all the going over.

And of course, because the garden is in some fundamental way 'ours', and reflects us, I am thinking of us as sixty-somethings, also going over. My own peculiarity of temperament means that lines and images from literature are never far, so, as we move about in the borders, I have in my head both the echo of Shakespeare's Sonnet 73, 'That time of year thou mayst in me behold' and Yeats's 'had pretty plumage once'.

But still, as well as the dying, there is abundance.

⌒

Nothing says summer like a blaze of zinnias. Ever since childhood I've loved these symmetrical, outrageously coloured, papier mâché-like flowers. The August gardens of the north-east of North America where I grew up are full of sweetcorn and tomatoes and zinnias. Buckets of them. Sometimes a gardener turns away from the commonplace, run-of-the-mill plant and cherishes instead the exotic as something to aspire to.

I confess I do like meconopsis – the blue Himalayan poppy, which is notoriously challenging to grow from seed, and bought plants never survive from one Kiltumper garden season to the next. A horticulturalist, Bill Terry, renowned for a personal journey with the blue poppy, wrote in his book, Blue Heaven: Encounters with the Blue Poppy, *that at the Chelsea Flower Show in 1927 garden lovers were falling over each other to snap up seedlings for a guinea apiece – $50 in today's market. Buyers wanted the plant as a trophy for their gardens. He* wrote: 'Look at it this way: had the Blue Poppy turned out to be the perfect plant, easy to grow, obligingly long-lived, readily self-seeding it would have become just another commonplace garden poppy, as familiar as the opium (Papaver somniferum) or the oriental (Papaver orientale). Definitely fetching, but unremarkable.'

Native to grasslands of the south-west of the US and Mexico and South America, it is unusual to have zinnias growing in Kiltumper, but each year I try. I have to grow them under cover of the front porch or in the tunnel and glasshouse. No way along this wild Atlantic way could they survive in the flower border. The Aztecs originally called them 'plants that are hard on the eyes' because of their colourful flowers. Even though they were originally brought to Europe by the Spaniards in the 1500s, they are so named because of a German botany professor, Johann Gottfried Zinn, who wrote the first description of the plant. There are twenty known species.

I love this snippet from the Chicago Botanic Garden site: 'Zinnias come in a preposterous palette (except blues) of every bright and pastel, plus bi-colours, tri-colours, and crazy quilt mixes.' There's one called 'Benary's Giant', which is three feet tall with sturdy stems and large flowers. The 'Zahara' variety is resistant to the scourge of almost all zinnias – powdery mildew. But the flowers have a delightful redeeming quality: even if their leaves are affected by mildew, the flowers are OK. The plants can be powdered in mildew but still covered in flowers. I spray the mixture of sodium bicarbonate and soap on them to halt the progression of the fungus and it helps a good bit.

The pots in which I once again tried to germinate the blue poppy seeds sit vacant of growth. No success, but in the glasshouse and tunnel I counted fifteen zinnia plants blazing away. Where blue poppies fail, zinnias succeed.

It's clear enough I am not an Aztec because for me their colourfulness is NOT hard on the eyes. Rather, it fills my soul with cheerfulness, and, God knows, every ounce of that I can resonate with does me a world of good.

~~~~~

In the tunnel, the courgettes are coming faster than we can eat them. The bounty chastens us. Why did we grow so many? *What, did you doubt that I would produce?* each plant asks. I am especially guilty in this. In the quickening pulse of spring I always think we should put in more than we need, that way if some fail, we will still have plenty. But of course, Chris makes sure none do fail. The courgettes, that drank much of the harvested rainwater, that seemed slow to come, and then suffered a yellowing of their leaves, catching a virus – *'They're not happy, Niall'* – Chris making up her spray of bread soda and soap, crawling under the large leaves to administer it upside down – whose first fruits came like half-inflated long balloons, bulbous at one end – finally took their medicine, settled themselves to their conditions and, well, the new fruits flourished. As anyone who has ever grown one knows, there is a battle with courgettes during August. You don't want to let them get too big. And each day there is surveillance and assessment, and each following day the plant defeats you. You looked away, and now look! They grow enormous overnight it seems, and while you carry in an armful for cooking, and giving away, you tell yourself, remember this now, *Next spring I'll follow C's lead and won't plant so many. I will trust.*

The spinach too is coming fast now, wanting to bolt. I understand bolting to be the plant's awareness that it is coming to an end time and that it tries to go to seed, to provide for the next year, and there is something so beneficent, so *thought-out*, if I can use that, in that it always gives me pause. I know the proper thing to do is to cut all the bolting shoots, and the spinach will likely throw out new leaves, but in the sentimental mood I fall into, I am slow to do it, and then it is too late.

The shallots are ready, they have done well in the dry ground and Chris declares herself happy with them this year. The beans, somehow elegant, and yes French, as they dangle beneath the leaves are plentiful.

But plenty is its own challenge in August, and Pythagoras takes a blow. The whole of the construction is a mass now of four different kinds of peas. None of the fence, neither canes nor twine, can be seen any more. It is just a green wall of abundance, which is fantastic, until the storms come. Rain belts down. We move into a time of Irish monsoons, for which there is no Irish word, I think the closest is *piobain* for downpour, but the pip-pip sound of that is harmless enough. We get wild rushes of wind, big-cheeked, forceful, and because everything is heavy with leaf, damaging. Pythagoras is a sail, and one morning takes a swing to the east. He doesn't go completely over, but he lifts his feet on the western side, and looks like a great elder on one leg daring falling, and not falling, at the same time. It is a triangle no longer isosceles, heading towards scalene, words I probably haven't used since secondary school, but am glad to have them cross my mind as together we press the feet back down. But the soil is loosened and nothing will hold the canes fully now. The fence wants to go over, is what you understand. The leaves of the peas are yellowed and in decline, but there is still a large crop.

'Should we just pick everything?' I ask.

'They're not finished growing.'

I can't argue with that. And so instead, as this mad August wind blows around us and the sail sways and is steadied and sways once more, we put a hook in the timber of the nearest raised bed, and then Chris runs green twine back and forth between it and the fence, watching the wind tension the line, until old Pythagoras is more or less upright. More or less isosceles. For today at least.

∼≺

From harvest to harvest, each summer we learn a little bit more. It's only our second year of the tunnel and we have much to learn, especially about ventilation. We tended to close the door but that makes an ideal climate for fungus. I asked Dr Garden Google and found a site that said baking soda (bicarbonate of soda) mixed with liquid soap – to help 'carry' it – and water would help the grey mould growing on the courgette leaves. The butternut squash we planted last year, inside the tunnel, has self-seeded in many places. The things you learn! The baby heart-shaped leaves are nearly too sweet to dig out, but we have no more room and they are robust growers, although the good thing about them is they are not as water hungry as the courgettes. I let one grow and am training it up a wooden trellis. For every success there is a failure but in the final analysis what we get out of the glasshouse and tunnel and in the few raised beds can't be measured in quantity. The joy on N's face as I return to the kitchen with a handful of peas and beans and cherry tomatoes, yellow and orange and red, and sprigs of basil and chives is priceless. We cook some brown rice pasta and feast on our achievements.

I'm turning Himself into a better gardener. Maybe he can turn me into a better writer…

∼≺

While it is impossible not to hear the whispers of autumn in the garden these days, there is still plenty in bloom. The agapanthus, both white and blue, are playing in the wind. Blue is such a rarity in the garden that the blue ones easily take your attention, but this year – maybe because we have had actual blue skies this summer – it is the white ones I find compelling. One white agapanthus just outside the front door has fifty, yes exactly fifty, I counted, arcing stems rising from the sword leaves, each one loaded with the fireworks of flower. I have little enough Greek, only some leakage that must have come across from five years of Latin, but I know that *agape* is love, and *anthos* flower, and to consider that, while looking at the plant, takes you into a dream place. And in that place, it's not that hard to imagine the person who came up with that name, what life experience or knowledge of love they had. As I said, it's a dreaming place, and that'll do fine for me.

Besides the agapanthus, the Japanese anemones along the cabin walls are in their heyday. A blue anchusa, a favourite, from the seeds Chris got when she went to the Flower Show last year, nursed the seedlings through the winter and then planted out in spring, is still doing well. There is a little lift of welcome in your heart when you see it has worked. An anchusa in August is a lovely thing, delicate, the branching arms of the stems like a botanical drawing, this stem here, this next exactly where you would put it if you were its creator, the flowers symmetrical and absolute sapphire. Because it comes in late summer, when much is in decline, it holds centre stage in the front bed just across from the table where we have our gardening mugs of tea, and I find looking at it lends me some serenity. Whether from the blue or the branching or Chris's success at planting the seeds, I can't say.

There is much that is dramatic in the garden at the moment, a tall pink phlox in one bed, a deep purple one in another; yellow, black-eyed rudbeckia; the orange daisy-like helenium that I learn is one of the commonly called sneezeweeds, whose dried leaves were powdered by the early Americans to make snuff. Learning that, I realise I know only the very fringes of the history of the plants we are growing, and resolve next year to do better, for this is another element of a garden, a gift of story and history, and I tell myself not to be only preoccupied with the sensual.

There is another yellow helianthus, a cousin of the sneezeweed, I think, and there are dahlias of course, still flowering in firework colours, and still cosmos and sweet peas that Chris grew from seed. She lined the back wall of the inside of the glasshouse with sweet peas.

'Won't they shade the tomatoes?' I had asked before thinking.

'No, Niall, it's on the north side. The sun shines from the south.'

The delphinium makes a last go at coming again, as does the French rose over by the cabins, blooming better the second time around for whatever the reason. Whatever it is, it's welcome. We have the blue-purple spire of the eryngium. I struggle a little with the name, some plants you can never remember, and to try and secure it I look it up in the RHS *Dictionary of Gardening* where I read: '*The flowers are small, packed in a cone-like hemispherical to cylindrical head, subtended by a ruff of spiny to lacerate, decorative bracts.*' And who could not love that? A ruff of spiny bracts. It's fantastic. To a writer, almost better than the plant, and certainly an added element of just sheer pleasure. The language attaches itself now to the eryngium, and for me at least does secure it better.

Also this: a thistle-like echinops – I learn that *echinos* is Greek for hedgehog, and looking at the spiky globe heads it makes immediate sense, and there is that moment I love again when the accuracy of the word brings you instantly across centuries to the one who first coined it, in this case with a little humour, and you and they are both looking at it with the one mind, hedgehog. There's something in that too.

The Swiss chard with its ruby stalks I am also counting in August's flowers. The RHS tells me it is 'sometimes grown as a garden ornamental'.

Well yes indeed, like Chris and I, garden ornamentals.

And of course, the roses are still blooming. All in all, even going over, the garden has a kind of glory. And when our neighbour Pauline's brother, Fintan, comes to help me cut up some logs, he steps into the garden through the open cabin and stops short to take a look.

'Well,' he says after a moment, 'isn't it just a little thrush's nest of a place?'

*

We get word late in the evening: starting at 6 a.m. the cement lorries will be coming to pour the base of the first turbine. They will be coming one after the other throughout the day until late afternoon. Because the excavated cavity for each turbine is large and deep, and because the cement mustn't begin to dry before all of it is poured, requiring dozens of trips out from Ennis, thirty kilometres away, we know this will be a fraught day. I wrestle with myself over this, *It is just one day after all. What is your problem? Get over yourself, this is the real world, you are a dreamer,* and go around and around a series of self-admonishments and accusations. But still, I can't quite escape a seeping sadness. What settles in the heart of me are not the old arguments about the 'green' value of so many lorries of concrete trundling back and forth all day on the sixty-kilometre round trips, not the burning of the diesel, the wear of the tyres, not the sinkage that will occur on both sides of the old Kiltumper road under the persistence of such weighty traffic (as the cement trucks try to keep to the edge of a road not wide enough for two cars to pass, the surface will begin to subside along the sides, the road will crack, fray, blister and pothole, starting to become bow-shaped by the following day); not the knowledge that the entire road will have to be remade, with more lorries, more stones, more tar and diesel, in a calculation against green values that is never made, not the noise, not the fact that 'green' energy is a clever misappropriation of the word, that in truth green energy is mostly grey, the colour of steel and concrete, no, it is none of these, it is instead the inescapable image of all that concrete filling two vast holes in what was a green hillside beside us.

Now, I know, *I know.*

I too hear all those voices about embracing the future, saving the planet, and all the rest of it. But I am grown old and wary and can't escape the feeling that mostly this is about money. Nothing else. And the truth is, that for this person – and maybe only this one, I concede – the *unnaturalness* of it, in such a green and natural place, is what fills the holes in me with sadness.

At 6:15 I wake with the first concrete mixer passing outside. The second one passes minutes later.

When I go downstairs, from the kitchen window I can see them across our meadow labouring up the hill to the site. Before the kettle has boiled, a third lorry passes outside. It will be a difficult day, a challenge either to work in the garden or to sit and write, because, in one form or another, it will be coming in on us. And be inescapable. If we had offices to go to, I think, not for the first time, it would be easier. But this house and garden is our office and today must be closed. By 9 a.m. I have stopped counting the lorries. I have Bach playing loudly throughout the house to try and screen the noise. Thirty minutes later, Chris and I get in the car and leave, edging out carefully between one cement mixer and the next, passing the man in a high-visibility vest directing the deliveries at the gate to the site next door, and leaving Kiltumper, not to return until that hole we can see in the hillside has been filled with concrete.

The following morning, an absence I can't put my finger on. The cement trucks are gone and the quiet reassembled. It is the quiet of our habitat, I think, the one we have been living in for so long we are inseparable from it. But in it today something else. I take my tea along the gravel path looking for what it is that is not there. I go past the tunnel, around by the well and neither find anything to explain this sense of something gone, nor do I escape it. I come back up to the house and stand a while looking out over the garden. Still, there it is, some vacancy. It takes me another ten minutes or so, during which I am aware that if someone were to stop by and ask me what I was doing, my answer of looking for what it is that is gone would only further an impression of oddness. Still, that was the truth. I stood and listened and looked and saw and heard precisely nothing that was unusual.

And then I realised what it was.

I went into the open cabin and looked up into rafters. I could see the nest, but no activity. Directly below it on the flag, the grey and white marking of the bird droppings was a day old.

That was it.

The swallows were gone.

*

From America comes a photograph of the front window of Dubray Books in Grafton Street in Dublin. It is filled with copies of my new novel, *This is Happiness*, published today. Deeply heartening, not so much in the promotion of my book, about which, like at my birthdays, I always feel queasy. But first, in the simple fact of bookshops themselves still existing in this afterlife long since their death was announced, second, in the idea of the image going from Dublin to America and back to us here, like an emissary bearing news that was secretly signifying, not so much any triumph but just the fact of carrying on, of still writing, still dreaming, still hoping to make one good book, and lastly, that we should see that image of the shop window in Dublin here, in Kiltumper, in the same room where those pages were written. It feels circular, right, and in some way, whole.

Japanese Anenome

9

September

The sycamore leaves are turning brown.

At first, it is just the leaves on a single branch, high up on a tree fifty feet tall on the eastern side of the garden. On this single branch the twenty or so leaves I count are a startling yellow, the rest of the tree green. Because we are still technically in August, because the heart of the summer was so warm and, for Kiltumper, dry, the sight of the leaves stops me a little today. And I think that, although I have lived now more than half of my life in this garden, still it can surprise me. If you asked me before today, I would have said the leaves of a tree turn gradually, and more or less all at the same time. So the sight of the single yellow branch high up there seems a kind of saying, and coming and going below it, every so often I glance up and hear: the year is turning.

*

In my notebook, this: *It is likely inevitable, those living on the margins forget the centre, as it has forgotten them. The margin becomes the centre of a different entity, a portion of which is unmapped and invisible.*

And back to this again: *As surely as our thirty-four years here have shaped this portion of Kiltumper, so too over the same time it has shaped us. This is something I am less aware of, having the necessary illusion of always being me, and C always being C, and, like the garden, unable to see the previous incarnations. The garden shows no sign that once a path ran down the centre from front door to the small iron gate that Tommy the blacksmith made in the forge in the village. It has forgotten the perennial borders we put in either side, and the giant rhubarb that grew out of a dung heap by the front door, the holly tree over on the eastern wall that was stripped of lower branches each December by one and all who came calling for a sprig with berries. So too our earlier selves here are as unreal as photographs and make sense only as part of a growing to become this, today's iteration, the people shaped by this place as we were trying to shape it. A garden always seems to have been growing to this moment. A garden is always now.*

*

I'm not sure if it happens everywhere in the west, but hereabouts people say, 'You're promised a fine day.' It's a verb I like. *Promised.* The pledge, assurance and the presage in it. 'You're promised a fine day, Tuesday.' On the surface, the promise might be coming from the weather forecaster, but there's something deeper and older in it, something of both an unwritten and still unbroken covenant between humanity and the maker of weather, and the optimism that a promise is an undertaking that will be kept. That, as often as not, in these parts, that promise is not, or not quite kept, has not eroded the usage, and I have heard young people as well as old saying it. And people are forgiving. When the promised fine day doesn't actually come on Tuesday, you'll hear, 'You're promised it now on the weekend.' The promise flies on ahead, and not sooner but later, in an uncalendared time of its own choosing, there it is, the fine day. *Didn't I tell you?*

We have grown used to this – it probably only took the first ten years of fine days not arriving when they were expected, coming of their own accord, singly, in clusters, or not, it seemed, for weeks – and so when the promised sun and heatwave of this weekend went elsewhere and nothing but a fine drizzle settled over Kiltumper, we didn't do more than shrug and say a silent '*heatwave?*' to each other. It has been a good summer, by all standards except the one of last summer. The thing is, it is not just the missing out of a couple of sunny days that mattered, it is the sense that settles silently over both of us in the drizzle as we come and go in the garden today. It is this: the summer is over. Already, I am aware of myself noticing the light in the evenings diminishing. The moment you notice it there is that little fall in your heart, and in my case at least, the realisation that I didn't appreciate how marvellous, how actually wonder-filled has been the brightness we have been living in, the lit evenings past eleven. It is the same each year. Each year, I can't wait for the light to come in the spring, tell myself I will treasure it, because who knows how many more summers, how many more long gleaming evenings we will both have, and by the start of August then I have somehow become used to it, it becomes our habitat, and I don't acknowledge it each day. I accept it. And only realise that I have when, in early September, the light is already leaving. A week of poor weather, and when the promised fine day finally arrives you notice with a little shock how much has been trimmed off the daylight in the evening.

There is something of this in the plants and vegetables too, and Chris has an especial urgency these days to try and get the last bit of growth and bloom out of them. She is working against the same diminishment I feel in the light and the coolness coming in the air. The garden is still so full of colour that it is hard to picture the bare beds of winter, but Chris knows each plant and each part of their cycles so that when I am working alongside her like today, the drizzle a kind of day's breath misting over us, I know what I am witnessing is an intimate relationship between a woman and the earth. Earth again with a small 'e'. This piece here, that her great-grandfather worked. It is a thing almost beyond telling, I think, for it matters to us in the most personal way and is at the heart of the way we have lived and continue to live. From it radiates the sense that this is something worth doing. Essentially, that there is some human worth in it.

We work in the drizzle all the promised dry day, until I am somewhere past exhausted and need to take a nap, and Chris takes a glass of red wine around some 'small jobs' she still needs to see to. Unfortunately for her she is not a napper, no matter how hard she tries. 'Triple Aries, remember,' she says. 'Fire never sleeps.'

Later, during a dinner harvested from the tunnel, the drizzle lifts and a lance of late sunlight comes right across the garden from the west. It spotlights the yellow faces of the helianthus in the lower bed and is so strong it lends them an unreal look. You think you've never seen them so lit, and then you look up and see that, somehow, remarkably, here, at the very end of it, in the last chance saloon, is the promised fine day. The sky is just entirely blue.

Now, it is often said, and by both of us too, that one fine day in the west of Ireland is worth a handful elsewhere. Well, one hour of this golden evening light is greater than its measure too. Rarity is part of beauty, and we are not too old – or maybe now just old enough – to exclaim aloud 'Look at the light!' and drop everything and go to the front window or into the conservatory to see how the late sun has entered the garden from the west and how it holds it now. And maybe because of that rarity, and because we have worked in the garden all day and know how each growing thing in it is hungering for this sunlight, the moment is a promise, kept.

The full moon falls late on Friday the thirteenth. Because it rises closest to the autumn equinox September's moon is also the Harvest Moon – so named because there is still light for farmers to continue to gather their harvest. (Moon-watching is a thing I do.) And on Friday the thirteenth, well, I'm feeling a bit like a lunatic. Day and night are nearly in equal measure. I'm out and about until nine in the evening with my companion glass of wine, pottering.

September is a month of contrasts, with summer lingering and autumn either officially having arrived (1 September), or approaching (23 September), depending if you follow the meteorologists or the astronomers. (There is also another way of telling when autumn starts, which is used by phenologists – scientists who study plant life cycles in relation to the changes in season and climate. But I can't tell what the plants are saying in this drastic climate of change.)

Leaves are twirling in the cooling air, yet tomatoes ripen still in the heat of the glasshouse. Lettuce seedlings are growing in the tunnel and the butternuts are bulging. The yellow zucchini from the seeds I bought in New York last April are tiny compared with their green cousins but so sweet. Butterflies are gorging on the sedum and on Michaelmas daisies. At any one time there can be thirty to forty peacock and small tortoiseshell butterflies and maybe even a red admiral fluttering wings in some silent music, savouring.

I can barely see the moon rising, tonight, between the clouds. But if I counted the number of full-moon risings I have watched it would well exceed the twenty that Paul Bowles refers to when he wrote in The Sheltering Sky: *'How many more times will you watch the full moon rise? Perhaps twenty. And yet it all seems limitless.' He's talking about loneliness and the illusion of time and what the night sky shelters us from. When I read that around the time of his death in 1999, a year in which my own mother had died six months previously, I determined to make a point of full-moon-watching. I wonder if he knew that in that final year before the millennium it was a year of two blue moons. It only happens every eighteen years. Did he get to watch all fourteen moons rise from his home in Morocco?*

Nearly twenty years ago we travelled around the world, more or less, for nine months with our children and I saw the moon rise from nine different countries. It was a lucky year in many ways, one with our children 24/7, one without winter and without clouds obscuring the full moon.

Seven years later, when my beloved brother Stephen died, we gathered that night in the hills above San Francisco, his four children and their mother, my sister and brothers and our father, and we howled at the rising moon known as the Full Flower Moon, or the Milk Moon, or the Full Corn Planting Moon and said goodbye. By the end of that month there would be a Blue Moon.

Moons have meaning in my life. Tonight, though I can barely see where it is hiding in the clouded sky, I head down the road and up the hill past Martin and Pauline's. There, on the top of the rise, I stand sentry and watch. I whisper hello to those who have died and say, 'Come on, Moon!'

～

Feeling low, and somewhat defeated at the end of a day trying to write something, I needed to get out and walk. At the moment – I don't know about the future – Chris doesn't want to walk our usual walk west because it takes us past the site of the turbines, the metal gates and fencing that have arrived, the haphazard mounds that look like a demolition site of gravel and stones where once there was a centuries-old stone wall, the general clearing and roadworks of any industrial site, and in this place, which by nature is the very opposite of all of that, it upsets her. It is the road we have walked for decades in all phases of our adult life; it has become part of the landscape of us, and I can understand why now she doesn't want to walk it. She prefers to stay inside the hedge line of the garden or walk the opposite way to the Blessed Well.

But this evening I needed to walk, for the oldest reason: to get past the place I was at.

It's a strange thing, with the whole of the countryside there before you, that you can feel confined, but what can I say?

I walked the road west in late-evening sunlight. The lorries and the diggers were gone and the natural quiet had reassembled for now, and the birds came back. The stillness felt strange almost, and I realised two things, how full of noise and movement the days have become since the construction began, and also that for thirty-four years it has been this quiet, but I have grown used to it. This quiet has been our habitat. I have noticed it often when I find myself in a crowded restaurant in a city and marvel at how people are able to hear each other amidst all the talk when I cannot.

The other thing that entered me was the sour knowledge that now that level of hush and stillness in Kiltumper will be no more. I had never quite grasped that. I had never thought that one day this road would not have this cloak of green silence, but instead have the whoosh clockwork of the metal blades, with subsequent shudder-pulse travelling down into the bog and the water table and down the hillside of Tumper. An engineered and inescapable constancy that no matter what the PR people of the energy companies claim, will not be as silent as this walk this evening.

Thinking that, I understood that, to many, and certainly to developers, and to *An Bord Pleanála*, the government body who judges objections, quietness has small value, so we are alone with the sense of loss and the sinking heart that accompanies the too-late realisation of something precious passing away.

'Still, we are both still alive,' I said out loud to the dip in the road by the old quarry. 'We are here now.'

I was trying to argue myself into the present moment, which I presume is not the recommended way to serenity.

I came up the rise past Charlie's – past the three hazel trees, sisters, that Chris has spotted growing on the northside ditch (*'I'd love to coppice those. Look how straight they are, they'd make an excellent pea fence.'*). In this dip and rise of no more than a hundred metres there is a different climate: if it is windy along the road it is not windy here; if there is frost, even the times when there have been dustings of snow, neither are here, a place on the road with its own character and strangeness, where in olden times they might have attached fairy lore – well, I came up there this evening and what I came up into was light: the sun going down was stronger, brighter, fiercer, than in its rising or all through the day when screened by cloud.

It was a stop-and-look moment. One of those moments that can occur in the west of Ireland when your brain is telling you this is a John Hinde postcard, this can't be real, and at the same time, you are there looking at it, the green sweep of the valley, the repose, the line of the land running towards the horizon where the ocean is a pale gleam. It *is* real.

And this is the thing I constantly have to remind myself. We live in this place somewhere between the idyllic and the real. And both are true. Along the road all of the ferns wore grey petticoats of mud from the passing cement lorries. The uncustomary traffic daily leaves its own mark, a crisp bag, thrown out a window and garish in the ditch, a plastic Lucozade bottle, further along a Wispa chocolate wrapper, the foil shining foreign in the grass. I picked them up and brought them with me, not from virtue, but because they hurt my love of the place. It's the truth. So that too was the real. But then there, in that stillness and remarkable light, came travelling a scent out of all late summers, the honey of meadowsweet. I had to follow my nose before I saw and translated the scent releasing in the late sun after a day of rain. And when I looked along the hedgerow, meadowsweet was all I could see; the delicate clustered pale flowers of it were everywhere in the wild places that have not been cleared by the diggers to widen the road. And it was simply beautiful. So much of what I notice in the natural world I notice *because Chris noticed it*, I thought. Because she cannot walk the road without seeing a wild primrose or a twist of honeysuckle or a foxglove that would do better if she just went in and pushed back that strand of bramble. She cannot help herself. And this evening, walking without her, I was aware both of how much I have learned, and seen, through my proximity, as it were, to her way of being, and also that, since I turned sixty, and have a greater general awareness of approaching an ending, however near or far that is, I have been making a more conscious effort to be a better noticer. (*So much I have not noticed. So much I have taken for granted. But it is not too late*, is part of a more or less constant internal dialogue.) In seeing the meadowsweet on my own, I was already looking forward to telling Chris about it when I got back.

Now, maybe all this is just to say, human beings are no different to plants who, we have learned, prosper and flourish depending on the company they keep. In any case, the fact was, the meadowsweet lifted me. I remember reading of the ancient Irish idea that there were 365 herbs, and 365 parts of the body, one for each. I knew that the English name derived from its use in sweetening mead, that it was strewn on the floor of cabins to make them smell sweeter, was used as a cure for a number of ailments but I could not remember which exactly. It didn't matter. The meadowsweet had been there every year, every September rising out of the green miry growth along the walls, but this evening, caught in the light, and sweetening the air along that place where the tongue of the road was out and titled upwards, it met some sorrowing in me, and well, cured it.

I know that might seem odd. Still, so it was.

I plucked a sprig for Chris; and then, one hand full of the lorry drivers' plastic bags and rubbish, the other holding the meadowsweet, I came back along the road, meeting no one at all, but with the light on my shoulders. When I came in through the open cabin the garden was bathed in such generous light that my feeling echoed the words of Henry Mitchell: 'Almost any garden, if you see it at just the right moment, can be confused with paradise.'

*

We have two large blueberry bushes, and each year the blueberries come, and each year, about two weeks before they will be ready for picking, we set up a large net over them. The net was one we used in a school production of *Carousel* years ago, I can't remember exactly how, but it has had a good afterlife in the garden. It goes over the bushes, and around the base we put four logs to keep it in place. The bushes at this stage are each over five feet tall and four wide, and look fairly unsightly wearing the net. I don't like the look of it one bit, but the blueberries have the promise of being more delicious than any you can buy, and will be the only blueberries grown in Kiltumper, and the thought of that makes them even juicier.

Previously, every year, as though by a bird-telegraph that announced *The berries are perfectly ripe now!* I would come outside and find the bushes completely nude of berries. The birds would come in the dawn and strip them in a single go. The sight of the bushes with the stamens still attached was sort of astonishing, and between the marvel of such a feasting and the awareness you wouldn't be eating any blueberries that year you were trapped.

This summer the dry and warm patch of weather meant that the berries-in-mind were the most delicious ever tasted. (There must be this in all growers, the anticipation is a large part of the pleasure. Everything that grows peaks for so short a time that the longer time of waiting and hoping has to be a key part of the whole experience – there is a version of this in the writing life too, when all the time you are thinking of the book that is still only in your head, and no sooner is it in your hand than you are thinking of the next one.)

Well, yes, these were going to be proper blueberries this year. We watched them fatten with the heat, plucked an occasional few – under the watchful eyes of the birds no doubt – but they were not ready yet. At the end of June I set the *Carousel* net and determined that this year, finally, I would beat the birds to the berries by picking them all *three days before they were at their peak*, that moment when between thumb and forefinger the berry, with spirit-moving generosity, *gives itself to you*.

Now, as I've said, most mornings I am generally writing, and it is the afternoon when I go out to work. That's the shape of most of our days. Chris gets outside before I do and often, seeing her, I have to draw the mental curtain that keeps me inside a fictional world until the afternoon. When she writes she often writes in the evening, and can concentrate better than anyone I know. Hours can pass and she can be clicking away on the keyboard, oblivious to everything but the rising of the moon. I know I should try and write in the evenings too; I could get more done outside, but *two* writers writing at the same time…

Well, the day when I knew we were going to pick the blueberries I went out before starting writing. I went out just to breathe in the light and the air already warm. I love the stillness of early morning in the garden when nothing is happening in it on a human scale, and everything on a natural one. You feel like a privileged visitor and try to move quieter than human beings can.

I went across the front of the house and along the pebble path and before I got to the blueberry bushes, I could hear the flutter of wings.

The bushes were alive with thrumming.

I looked inside the net and saw two things at the same time. The first was that nearly every berry was gone, the second that there were two blackbirds trapped inside the net frantically fluttering about, flying, falling, flying again. The literal bird-in-a-bush, times two.

Another two things occurred to me. The first: the berries had not yet been at their peak. They had been still a few days off. How did the birds know I was going to take them all? The second: what a job they had done! If you hired them for picking, you'd rehire them the following year. They had *cleaned* it. I imagined a dawn feasting, the bushes surrounded by fluttering wings like something out of a garden mythology.

I could be disappointed, but not angry. The berries had not gone to waste, and *they had been delicious* I told myself, just not for us.

I lifted the net. The birds didn't fly out at once. Maybe they had stopped believing in escape. Maybe they were thinking the feast worth it. Maybe they were drunk on the superfood of blueberries. Whichever it was, it took some coaxing to get the thieves out. When at last they took off, berry-full, they only flew across to the old sycamore twenty feet away and perched there where I could see them as I took down the net.

'Well, is your song improved?' I wanted to ask them, but there was no singing for now. Imperatives of digestion, I suppose, taking precedence.

Well, that was back in August. But I thought of it today when I was pushing a barrow of Chris's cuttings past the bushes. Because I was arrested by the superb autumn colour. Here in Kiltumper the turning of autumn tends towards bronze and yellow. There are few enough things that turn red, but the rich tones of the blueberry leaves were something to be, well, savoured. You had to just stop and take them in. The words for the colours fell short of the experience, as words do sometimes in the immediate surprise of something truly beautiful. Not a thing a writer should easily cede perhaps, but true in my thinking. So, I just let rest the barrow and looked and *fed on* the blueberry bushes. I smiled and took a glance towards the invisible birds in the old sycamore and thought: *Ye didn't take this.*

And before I had taken up the barrow again, I knew it was worth growing the blueberries just for the glory of the leaves, for the blaze of their colour in the autumn.

Let the birds take the berries, we'll have the beauty, which requires no net.

*

I think Chris loves the garden because a garden is truthful.

*

Morning begins with the harsh noise of machinery. Rain is sweeping down in waves, a large lift is parked just outside the gate and a man in a high-visibility suit with a red helmet is pressing the button to have the machine bring him up to the top of the old ash tree. He has two chainsaws. The ash tree is well over fifty feet tall, is healthy and multi-limbed, and in full leaf. It is not one of the species of ash that are suffering from dieback. The seed established itself and grew by the ditch many decades ago – I like to imagine that moment of establishment, this seed among all the others that blew away and perished, this one surviving, finding adequate earth, humus in the untended tangle of the ditch, the first hair-like root system reaching out, secretly burrowing, drawing nourishment, in the race to literally implant itself before the first winter; the smallness of that action across

the road from this house, the world that was then, a time
before electricity, the sapling that was spared when someone
driving cattle in the gate looked to pluck a random rod, and
instead chose another; the first time anyone noticed it, the
first time it earned the name *tree*, when it had already been
alive and growing maybe five years or so, and its few feet did
not yet guarantee its survival; the tough ash-endurance in it,
the hardness, which it reminds you is necessary for all living
things in this place in the path of the Atlantic's last puff, and
yet, in the spread splay of its branches, softness and triumph,
the old tree's leaves still a vivid and pulsing green.

Left to its own devices, there on the side of the ditch, the
tree has flourished, and looked at now is probably the finest
of all the ash trees around. But it does tilt out towards the
road – I don't believe the angle has changed in all the years we
have been looking at it, and that's not the reason the tree has
to come down. It will interfere with the turbine blades on the
day they have to pass by. (How simple is that thinking? How
logical to developers.) The roads in the west of Ireland were
made to human dimension and needs, and with responsiveness
to the land's contours. None of them are straight. But the
turbines are not in these dimensions, and the transport of
them requires that the roads be widened and straightened so
that they can get into those rural places that, because they
are considered to have small or no scenic value, have been
designated suitable for wind development. Kiltumper is one
such. So our winding, tree-lined road must be straightened
and the old ash tree be the first to be cut down.

We have known this for a good while. Still, once the
chainsaw starts, the snarl of it is immediately raw. It will be
impossible to escape it today. It will be in every room of the
house all the rest of the day, for the tree will not go easily, it

will take hours; with all doors and windows closed we will still hear the snarl and moan of the saw interrupted only by the whoosh and crash as another great limb wings down and lands in the road. In the pouring rain, two silent fellows in neon jackets will pause any cars coming or going, they'll drag the branches, throw them up on the ditch. Because they are just outside our gate, I don't think it remarkable that I have come out to see, but perhaps because they are vaguely aware I am unhappy or, less likely perhaps, share the feeling the whole thing has an air of transgression against nature, they don't look in my direction as they haul off the fallen wood. It is a dour scene, dismal and damp.

I am aware there are some people who do not like trees. Some farmers who consider a tree a waste of good ground, others, especially older people, who consider trees dangerous, always on the point of falling on top of someone, and so, to these the chainsaw may be a happy sound. We are not among them. I myself hate all mechanised tools. We have a lawn mower and a hedge trimmer, and if I could get away without using either I would. Each year when the leaves come down – and it will be soon now – we spend several afternoons raking them, and each year our kind neighbour Martin, passing on his walk, pauses to offer me the use of his leaf blower; each year I decline. I just have an immediate aversion, both to the physical – noise, shudder, force, fumes, of machined tools in the garden – and the abstract, the idea of what they enable, none of which seems tilted towards the constructive. So, I may be the last person to appreciate the wonder of a skilled chainsaw man. He is high in the ash tree, three times the height of the stone cabin across the way. The saw eats into the timber. There's an air of man's implacable will, and yes, violence, about it, the morning ripped up and a constant snow of sawdust flying out over the road and into the garden.

I go inside. When I close the door the noise of the saw is still plain. I turn up the radio.

'Well?' Chris asks.

'He's a professional tree surgeon, he's doing a good job.' It's all I can offer.

'Is he leaving the two smaller trees behind the ash?'

'Yes.'

(Those two *will* survive the day, but in the morning, it will be decided they too have to go, and an hour's work will take them down.)

'I can't believe you can just take away a healthy ash tree that has been growing for decades,' Chris says. 'How many trees along the road are going to have to go?'

Most of them is the answer, but 'I don't know' is what I say. My spirit is exhausted from trying to find positives. The truth is that many mature trees on the Kiltumper road will be felled in the next two days, along with every branch that extends out over it and may interfere with the passing of the turbines. Again, the fact that this is being done solely for the one-day transit of the turbines leaves a sickening taste. It is not only the fact that Ireland has one of the lowest percentages of forest in Europe, that all government policy-speakers assure us they are addressing this. It is something visceral. For Chris, who has anger where I have disappointment, you can't just come in and sit down and have a cup of tea while the saw is moaning outside.

'On the websites of the wind energy companies, do they say they will cut down all the trees in their way?' she asks. 'For two turbines, in a place where there are ten houses, they can take away your trees?'

And where does it say that for every tree they cut down they will plant another one?

Answer: nowhere.

'It's not all the trees.'

'It's not the ones on our side of the road because we wouldn't let them.'

Her chin rises and the idea of the thing comes like a dark wave through her and she sits on the couch and cries. I sit beside her and say all the usual things people say, basically different versions of that it will be all right.

'I know. I know it's stupid. Just feels so … wrong,' she says.

Now, we may be the only two people sitting inside their house upset that some ash and sycamore trees along the road they've been walking for decades are being sawn down. I realise that, but it's the truth, we *are* upset. She is also aware that on the first day she arrived at the place where her Breens were born, one of the notable features of the place was the number of mature trees. There are many on this road, and more on our property than many other homes in our parish. As I've said, the field next to the garden, and now partly in it, is called, has always been called 'the Grove', because of her great-grandfather Jack, who grew trees. So, trees had, and have, a special place in the Kiltumper story, and in Chris's family's part in it, the specimen trees we planted in the front garden, others around the place, and Chris's father planted the twin fields of Upper and Lower Tumper with ash and oak trees (which, ironically, will now be under the shadow of the turbines). So, trees have a place in the story, and the taking away of them today is a melancholy chapter.

The rain pours. The chainsawing continues. Inescapable, whether inside or out.

It's only for a few days, N says, all this noise and distribution and disturbance. It'll end.

Will we ever go back to the music of the everyday of the countryside? Where we are largely on our own, often absolutely, where the only sound is the birds or farm animals. Or T in his tractor passing with two dogs in his wake, M in his tractor with his salute-wave, and T in his postman's van, the cars passing at two and three to collect the children from the school at the end of the road.

To this country music will soon be added the spinning blades chopping the air like giant helicopters. Whump Whump Whump...

My work is writing, the place I do it is the front room overlooking the garden, so out of habit and need for escape I go in there. Czeslaw Milosz says that the ideal occupation of a poet is the contemplation of the word 'is'. But the tree surgery is thirty feet away and in my notebook all I can do is list verbs in pursuit of the exact sound, *growl, groan, snarl, roar,* all of which I note belong to the bestial, but none of which capture the jagged edge of the air itself, or the sense of a serrated blade cutting into the soft stuff of your brain. In the pause while the surgeon moves to another tier of the tree, my ears ring. They will be ringing in the silence after, and the following days too, as a kind of sonar aftermath lingers. The other reason I can't concentrate is that I keep looking. The ash tree, which was already tall and full when we moved here, rose high above the bottom edge of stone cabin, so that from where I sit each day to write, it shapes the outer edge of the garden with a green canopy. Chris says it frames perfectly the end of the garden. Ash are the last trees to leaf, but also stay green the longest, and so the tree was both part of the clockwork of the year – we watched for it to green, after the sycamores – and part of the framework of the garden. It sheltered the garden from the prevailing wind from the south-west. But now, all the branches gone and the main trunk being mauled by a digger, from where I sat there was no tree left, instead a vacancy, and sky, and that was the thing I couldn't stop looking at. Seeing more sky is not a bad thing, and in time we know we will get used to it, and we know all the rest of those arguments and the things people say about change, and so on, to get you past the moment. And they are all true, we *will* get used to it and will grow to like it even. That's not the point I'm trying to make here. It is this: I would never have imagined, could never have, what I would feel looking at the absence where for more than half of my life that tree had been. The physical presence of it had become part of my known world, and really, I had taken it for granted, the same way I did the road or the cabins. It was just there, and only now, when it was gone, did I feel its loss as one of the fixtures of the place. I don't mean to suggest here that it was my favourite tree, a special or beloved one, or anything like that. No, it was so ordinary in fact, so much a part of the ordinary natural beauty of the place that it only became fully visible when it was no more.

I keep looking at the sky-gap where it was.

There is a much broader view of the valley and a great sweep all the way across three townlands, so that is something; yes, the winds will be coming from there in the winter, straight into the garden, and we will soon be planning how to deal with that. But for now, I can't see past the vacancy, and the failure in myself to have fully appreciated the tree that is now gone.

*

In the evening we get word that the road-widening will start tomorrow. Now that the trees are cleared, the stone walls all along the southern side of the road have to be bulldozed and an extra two metres added to the width of the road, from O'Shea's at one end to the midway point just past our gate where the iron fencing marks the entry site for the turbines.

'It'll be fairly full-on,' the message says. I appreciate the warning, and am caught between taking Chris away somewhere for the week while it is happening, and staying to protect the boundary on our side of the road.

Chris is pragmatic. 'We have to be here,' she says.

It begins at 7:30 a.m. Eyes opening to the shuddering sound of caterpillar tank wheels clattering down the road. You come downstairs and outside in time to see the digger take a maw of stone wall. It takes six feet in seconds, turns and drops the mouthful three metres inside in the opposite field. On to the next piece. The stone wall across from the house starts disappearing before breakfast and with it the climbing rose that Chris was training over the wall and other plants she transplanted there from the garden. I am glad I saved Paddy Cotter's white fuchsia. Over a kilometre of stone wall goes. A cavity three metres wide and a metre deep will be dug out, and then the lorries will be coming with the rough gravel to fill it in. Then the diggers will pound that down, and it will be rolled until it is a solid foundation for the council to come and add the top-surface chip and tar.

Across from our garden the wall is more ditch than stonework. It was faced with stone once but grass and growth overtook it, and so in the years since we've been here we borrowed the tactic of Mary Breen, who lived here before us, and stuck cuttings and the divided plants that were in abundance into the verge, so that in summertime it became a wild flowering of everything from a rambling pink rose to montbretia to monkshood, an escaped delphinium, large-leafed pink geraniums and goldenrod.

Now, it all gets taken in the mouth of the digger. Essentially, in our minds, what is going today is history, not in the abstract but in the most physical and real way possible, in the actuality of the road itself and those who made it.

This is the thing that Chris speaks of most often these days. *'Is there no value on a stone wall? And the people who built it?'*

If developers come and say that in order for the ease of their business, we should straighten all the roads of the west of Ireland, we should demolish all the stone walls, would anyone stop them? The planners who approved the wind farm beside us knew that all the stone walls would have to go to make the road wide enough for the day the turbines would have to come in, but rather than say this is not a suitable site for industrial development, they said fine. It's a way of thinking with which I can't reconcile. We want one of the planners to be here today, watching what we are watching. We want any of the councillors who every five years come along the road and knock at the door to say they represent us, to be here. We want them to see and say yes this is exactly right. Because we feel it is exactly *wrong*.

The digger driver is skilled and quick and once he has planted the digger in on the levelled ditch, he can move along outside our hedge taking the wall and the rose and the Japanese anemones and the montbretia in no time. Which I think is precisely what this is, no-time. It is an unreal thing to see how quickly a handmade wall that a century ago must have taken weeks to build, that Chris reminds me

Jack-of-the-Grove and his people probably helped in the making when they were children, can be gone. Once, these stones were picked day after day, often by large families of children, to make the fields, stones brought to the perimeters for the building of the walls. Whole families were in the making. The stone walls hold the history of their hands, and though the walls are here and there off-kilter, here and there in tumble-style, or overgrown with blackberry brambles, they have the dignity of their makers. The walls are literally full of time and, as with the trees, when gone, are missed more than they were loved.

The digger moves along the road, leaving a bank of earth a metre and a half high alongside it. Those trees that were spared by the chainsaw now have half their roots spliced, and all along the bank there are tree roots sticking out like severed limbs. Chris holds her face in her hands when she sees them. She walks past the digger and dumpster men without a word. She has decided she won't say anything. But when she gets to where the beautiful moss-covered wall beneath the arching sycamores was, where now there is only a mound of earth and shorn roots sticking out obscenely, she turns to one of the workers and says, 'This was once a *beautiful* wall.'

He nods. ''Twas.'

*

Pointless Conversation I:

'You'll have a fine wide road when they're all done, Niall.'

'Yes. Almost like driving into Tesco's.'

Pointless Conversation II:

'You'll have a great road now.'

'Yes, people will be able to go really fast past our house now.'

Pointless Conversation III:

'I find them beautiful, the windmills. You'll get used to them. Didn't you get used to the pylons?'

'The pylons don't move.'

'What?'

'Try and walk past a spinning turbine and not look at it.'

'What?'

___

Agriculture in Ireland is responsible for a large portion of the country's carbon emissions, water pollution and loss of biodiversity. I read this in the new Brown Envelope Seeds (organic farm in west Cork) brochure. Six million cattle produce the equivalent amount of sewage in terms of biological oxygen demand as do a hundred million people, and produce methane, which is worse than carbon dioxide.

Once a wind farm goes in, farmers can put their cattle back in underneath it. Something is not right in this picture.

___

I am very aware that in all of this there are larger arguments, and the truth is that none of them are as easy or straightforward as the PR people and the politicians claim. The world does need to stop using fossil fuels, and wind energy has to be part of that solution. Chris and I would agree with that. Reading Bill McKibben, I learn that this summer began with the hottest June ever recorded, until July beat that and became the hottest ever. The UK, France and Germany, which together have the oldest recordings of weather in the world, all hit record temperatures this summer, 'And then the heat moved north, until most of Greenland was melting and immense Siberian wildfires were sending great clouds of carbon skyward' (*New Yorker*, 17 September 2019). Last week, Hurricane

Dorian laid waste to the Bahamas in what has been called the longest siege of violent, destructive weather ever observed on earth. So, yes, radical solutions are needed – in the article Bill McKibben calls for banks to stop funding fossil fuel companies, and if they don't, for people to stop using those banks. And the reality of the urgency of that need gives me pause, and soon enough I castigate myself for not embracing the turbines. What is some noise and inconvenience when measured against saving the planet?

Turns out I can castigate myself pretty well. I can get to a place where I am ungrateful and selfish and small-minded and narrow-sighted, where I am the villain who will not save the world. *Come on, put up your turbines*, is where this argument leads.

But then, something sticks; the reasoning has something awry in it. I can't put my finger on it right away. I am resolved that yes, we need wind energy, so is it only because one of the turbines is to be 500 metres from our bedroom window that I am upset? (*Surely, I can be bigger than that? Surely, I could take one for the planet?*) I keep turning it over in my mind to try and get at what is sticking. At the petrol pumps down at Fitzpatrick's in the village, I meet a man for whom any argument against the turbines is a nonsense.

'They are saving us for the future,' he says. 'They're getting us off the oil. We're already twenty per cent renewables.'

'Sustainable energy.'

'That's it.'

'Because the wind is free. Doesn't have to be drilled for, and pumped and refined and stored and loaded into tankers and brought all the way from Arabia, pumped out again, and transported all the way to these pumps here. Imagine the cost of all that, compared to wind. Which is right there, and free.'

'Exactly.'

'Instead of oil, free wind. That's why our electricity bills have gone down by twenty per cent.'

That gave him pause. For a single moment he thought it possible, then: 'Mine hasn't gone down. Has yours?'

'No.'

Which was unfair, I know. I couldn't resist. I could become a fanatic yet. It was just a kick against the story we've been sold. It's the same as when you ask someone to show you the carbon calculation of all the months of digging, site-clearing, tree-felling, quarrying, loading, cement-mixing, transporting, unloading, road-widening, construction, erection, connection to the grid, and ask them to add to that the making of the steel of the turbines, the blades, in a location hundreds of miles away, and tot onto that the transportation of the same in vehicles burning fossil fuels, and when you've done that carbon maths, show me the figure versus the green energy value of two turbines. And when they can't tell you that figure, you'll find yourself thinking this is all more about profit than climate.

And I'm not against profit, but I am against a false story.

So, finally to be done with this, let me say here that, tonight, still twisting on this, I think my position is this: I am in favour of wind energy. I am in favour of it, in the ocean. As is the case off the east coast of America, where they have put the turbines fifty-six kilometres out to sea, so they cannot be seen from the land. Now, as I have said elsewhere, I can already hear the voices that say, firstly, 'That will impact the habitat of fish,' to which the answer is almost certainly yes, but only in the vicinity, and sadly humans have to take priority, and secondly, 'That will cost too much,' to which I can only give the same answer: money is never the right answer if the question is the earth.

And, as already said, the simple fact is, no more earth is being made. And what of it we have already pock-marked with turbines cannot be lived on.

Chris and I are not Not-In-My-Back-Yard-ers, we don't want turbines in your back yard either.

And, finally-finally tonight, I realise that what all of this has done is brought me face-to-face with the real conundrum: *What is the countryside for?*

But that is too big a question for tonight.

*

A murmuration of starlings this afternoon. Not an enormous one, just a darned patch in the pale sky. In it something of a crowd coming excitedly from a football match, where all are of the one tribe. For a moment I watch, and envy it.

*

In the morning, the first news of the worldwide climate strike, mostly by young people, who are leaving schools and colleges today to march and mass in order to protest the inadequate response of their governments to the climate crisis. The strike begins in Pacific island nations like Kiribati and the Solomon Islands whose continued existence is threatened by rising sea levels. 'We are not sinking, we are fighting,' is the chant there. In Australia, where ever more soaring summer temperatures have become so 'normal' they have become sidebars of the news, and the warming seas have already killed half of the Great Barrier Reef, half a million young people walk out of class. The strike takes place in over 150 countries, and is widely reported all day. And it is both a chastening and hopeful moment for someone as old as me. There is that almost overwhelming sense of failure, the failure not only of every elected person, but also of those who elected them, the failure of all regulators, of all charged in whatever ways with stewarding industry, banking, business, all of us in fact, because there is a certain, shared, collective guilt of having taken the earth for granted.

(Now, 'taking the earth for granted' is a phrase I cringe to write. It has an air of sermon, and admonishment, and is one of those airy pronouncements that have no actual substance, the very opposite in fact of the way we are trying to live here, which I hope is all about substance. But I use the phrase because the very airiness of it seems to me to partly explain why the climate crisis has been allowed to get to the stage it is at this morning. The Earth, in the abstract, invisible goddess, is not the same thing as the soil and the water and the air. The figures we hear all the time, the temperatures, and the sea levels, the percentage of glaciers melting, remain at more than an arm's distance, and maybe what we ourselves can't feel, touch, see, is always taken for granted.)

So, there is some shame, and in the big picture, and the small one, here in our daily life, we feel that.

Our daughter, visiting from New York, picks up the bottle of washing-up liquid we use. It's the leading brand, bought without thinking.

'Dad!' she says.

'What?'

I first think it is the plastic bottle she is objecting to. ('I know,' I'm about to say, 'we are trying to use less plastic, if we drive the twenty-two kilometres to Ennis there is a refill-your-own-bottle place. We just have to *think* of it.') But it is not the plastic, it's what's written on the back she wants me to read.

There, in very small print, too small for someone like me whose eyes have seen sixty years of life, it says this product is 'harmful to aquatic life, with long lasting effects'. It says it right there, but in decades of buying and using it, I had actually never read that. And I might not have either if it hadn't been for her pointing it out. I tell her we will never buy it again, and we won't either. Small things, but still. Chastisement.

And hope. Because, although I have lived through many protest marches that brought no result – from the first one I ever went on, against the building of the Dublin Corporation offices on the site of the original Viking settlement on the banks of the river Liffey, to the ones against the weapons-of-mass-destruction Iraq war – this one feels transformative. Feels like it might actually be that hopeful thing that brings an end to stasis, a *movement*.

So today, while the protest marches go ahead around the world, in Kiltumper Chris and I work with a heightened sense of habitat and climate, and enforcing small changes for the better. In gardening there are still too many plastic pots, and although these all go into recycling, where you have to hope for and trust in best environmental practices, Chris says we can do better with reusing them. 'We can also start saving cardboard toilet rolls as planters, and do much better with saving seed. Think of all the seeds this garden produces each year,' she says.

And for a moment I do. A single poppy head produces hundreds upon hundreds of seeds. I have seen them spill over my hands, black-speckle the sweated creases of my palms that I have brushed free without a thought.

'We're going to do better at saving them,' Chris says. She has a number of brown envelopes, and we go to the perennial beds now with envelopes and secateurs. It's a small scale, quiet, autumnal act, of chastisement and hope too. Of doing better.

As we go, the garden reveals itself full of butterflies. There are enough of them on the sedums to be momentarily mistaken as blossoms of orange and black, that rise as one as we approach, and take off in freeform formless patterns. There are dozens of them. A stop-and-look moment. And because all butterflies are young, and because I hadn't noticed them at first, but now see them suddenly everywhere inside the hedge line, my spirit lifts with them, somehow briefly conjoining them with the young marchers out there today, and find in them – hope.

*

Word comes that the new novel is being well received. Since 15 August 1991, I have read no reviews of my work. That was the day my first play premiered on the main stage at the Abbey Theatre. The production had had a fraught time in rehearsal, and the opening night was the most painful experience I've had in my writing life. During the interval in our fine clothes Chris and I stood in the tiny VIP lounge which only fit us and the board members of the theatre. Not a single person on the board spoke to us, until the bell rang for the second act and one of them acknowledged me as the writer by turning to me and saying 'Well it's not *Hedda Gabler* is it?' and walking away. The following day the headline of the review in the *Irish Times* was 'Total and Embarrassing Waste'. The phrase 'of a bad idea' was added for the country edition. I was thirty-three years old and trying to take my first steps as a writer. So, since then, I haven't bought the paper and cannot bear to read anything about myself or my work. I am on my own on this path is a good summation, Chris the only one sharing it with me. I understand that that sharing is not easy. It doesn't bother me so much but she suffers when a book is struck down, or more often, ignored. Which has come to seem an integral not to say natural part of our secluded life here. So, today, when she comes to me in the garden and says, 'The book is getting great reviews,' it is for her I am happiest.

*

Four diggers now working on the Kiltumper road, clatter and bang all day, gravel and more gravel arriving just outside the hedge, and yet, here, in fabulous sunlight in the garden, that September light that Monty Don rightly says has elegance and clarity, Chris sits to draw the grace of a Japanese anemone. She is a picture of concentration and beauty. This too is the way we are living. And by this, I think, so much is resolved.

Cherry Tomatoes and Geraniums

10

October

The windows of what some people optimistically call 'the sun room' have these thin graceful arcs of wood outside against the glass that give the view a church-like air as I sit in here every day writing. When I first come in and sit down, the view from this church is almost entirely green. And at first glance, when I look up from the page, green is the only colour that strikes my eye, then the mostly grey of the sky that is in truth less grey than pallor, a kind of watered colour, one that feels abluted, if that's the right word, abluted is the sound of it anyway, a washed light that gleams too much to retain your gaze so it brings your eye back down to the green. The gutters are singing their small-throat songs of rain. The longer you look at the green the more colours you see. Today, the Japanese anemones seem everywhere pinking, and your eye now dances from those over by the stone cabins to those by the centre flagstone steps where once the giant rhubarb grew. *Just keep looking* is what I write in my notebook. *Everything is revelation.*

'Things reveal themselves passing away.' That line from Yeats keeps coming to mind, and these October days I have been thinking a lot of that. Revelation in dying. It's not morbid, it's hopeful, I think. A kind of insight into the true nature of things seems offered in this season, and although personally I am today not aware of my dying, I am of moving into the fall of my life, and so want to have revealed to me some greater understanding as I go on.

The kind of things you think in a green church with a grey sky.

In Kiltumper now we come into Samhain, the season of the spirit, and the ripening and dying of living things according to the Druid calendar. The season of mists and mellow fruitfulness, according to Mr Keats.

N, meanwhile, is as ever busy writing. The leaves blow down past where he sits in the conservatory. Do his characters think of getting out a rake? With dark skies returning, two hundred billion stars of the Milky Way are lighting up the sky. (The ancient Irish called it Bealach na Bó Finne, *the Way of the White Cow. And I like that, easy to imagine that great cow slowly crossing the heavens.) The sky holds such history and myth for my imagination. On a dark night you can see over two million light years away. Like the stars spiralling above in the Milky Way Galaxy, petals of pink and red roses lie in a whorl on the ground in front of the cottage. A somewhat dramatic show of amber sycamore leaves fall on the grass in such perfection you'd think someone had scattered them there by design. Elsewhere, the leaves of blueberry bushes and acers and salix turn brilliant crimson. Rosehips as large as crab apples wait to be feasted on by soon-to-be-winter-hungry robins. It's time for the great autumn clean-up, but instead I watch the falling leaves from the quiet of the house and pray for a great wind to tidy them away.*

Donald Culross Peattie wrote in An Almanac for Moderns, *'It is nearly impossible to be sad, even listless, on a blue and gold October day, when the leaves rain down, rain down, not on a harsh wind, but quietly on the tingling air.' Whereas September was a month of contrasts with summer lingering and winter approaching, autumn has turned the corner with certitude. One time, long ago, my father was speaking lines of a poem to me he was starting to write. 'October, teach me how to die,' he said. We were driving on an autumn day in a suburb of New York City along the Westside Highway, stippled with red, orange and yellow leaves. At the time, it seemed rather curious to me, just a teenager, and my father a Wall Street lawyer, but I have never forgotten it. And now that image comes back to me. October, a month of endings and my father nearly ninety.*

Outside, I notice an abundance of sloes. The berries are so full and of such a serene navy they seem out of an older, less perturbed time. I can't quite explain that. There's a kind of fortitude in them too.

Are there more berries this year? Or do I just see them more?

*

In the *New York Times*, Carl Zimmer writes 'the skies are emptying'. The number of birds is in serious decline. In the last fifty years North America has lost 2.9 billion birds. The number of starlings down by 89 million, blackbirds down 440 million, and grassland species down 717 million. 'Europe,' he writes, 'is experiencing a similar loss, also among common species.'

It is impossible to read this without a fall in your heart and sense of human failure. In the large picture, you know that birds, like bees, have carried no weight in issues of planning and development. I remember reading a planning proposal for wind turbines, and the bird survey that was mandatory. The hen harrier is an endangered species and a nesting pair found at the proposed site could result in planning being refused. I am too long in the world now not to have been surprised to read: 'No hen harriers were found on the site.' Birds, literally, have no voice, and this morning when I look at the number of species in Ireland now threatened with extinction – twice as many as seven years ago – it is hard not to feel despair at the constant erosion of the natural world. Of the curlew, there are only 138 pairs left. Twenty years ago, Ireland was one of the strongholds of Europe for the bird. But with wetlands drained for development their habitats have become fragmented, and the curlew is vanishing. The cuckoo is now on the 'orange', in decline, list.

And I do fall into a gloom, and go outside to work with a sense of the frailty of everything. I have a heightened alertness to what birds move from the grass to the trees and back again, and pause to watch them, as though to see if they know.

But, what to do? An apologetic air will help nothing. I come inside in what I call my gardening clothes – which were my ordinary clothes until by natural exhaustion they declined to garden wear – and bring up the website of Biodiversity Ireland. The site lists ten ways to help biodiversity, and birds in particular. The first of these is *Retain Hedgerows*. When I read this, I have to lift my head and pause because more than a kilometre of hedgerow has just been dug away to make way for the transit of the turbines. But ours has not. And we are able only to mind that piece of ground that we are charged with. So, yes, we will retain wild hedgerows. The second thing on the site is *Plant Native Trees and Shrubs*. It names hazel, willow, hawthorn, wild cherry, elder, and this we will do too. (Chris orders twenty-five trees before evening.) Other advice – like not draining fields, allowing the growth of wild flowers, allowing dandelions, leaving some meadowland unmown – is easily followed because, (sometimes out of laziness, or because we have no machinery, and because there are only the two of us) we are already doing it.

In none of this do I have any sense of heroics. Nor do I fool myself that that is the birds taken care of. It is not. As with all climate- and man-related destruction of the environment, land, water and air, the situation is too perilous for anyone to feel complacency, and I don't. But I suppose awareness is something, intention another. I have noticed that some portion of my notebook this year is taking on the air of what Seamus Heaney called 'ecological lament', and while there is an inevitable aptness in that, I don't mean to say it darkens every day. It doesn't, but the truth is it is there in the background, more or less always now.

I close the computer screen to go back out to the garden. At the back door, I had footed off my wellingtons, and through the window I can see them. On the rim of the standing one, a robin.

⟫⟫

Tomatoes are not the only thing still ripe for harvesting. We have plentiful runner beans, spinach, kale, onions, zucchini, yellow squash, butternut squash and lettuce. The corn, a favourite from my childhood, and which I have tried to grow from seed this year in the tunnel, is now five feet tall and producing the first hairy tufts that will hopefully become the cobs. ('Ears of corn' makes sudden sense when you see where the cobs start to appear, sticking out at the sides, a little comically. Sixty years and more

*of living before that moment of seeing it so clearly, of course they
are 'ears'.) Without the sunshine, the Kiltumper corn won't be
able to compete with the taste of American sweet corn, which gets
that sun-kissed sweetness, but today I think it was worth growing
them just to see the first ever corn grown in this townland.*

The tower of the first turbine is standing.

All of the parts having been brought to the site during
the night, it is assembled relatively quickly. It rises, tubular
and grey-white, on the top of the hill of Tumper 500 metres
behind us, above us. By evening, the oval-shaped cockpit is
added, but not the blades, so it stands immobile like a raised
fist. We look at it from the kitchen window and begin what
may be a slow process of accommodating it inside the view
we have been looking at for thirty-four years.

As is my natural instinct, I try and be optimistic. 'At least
it's not in front or beside us. It's not as tall as I thought it
would be.'

But Chris, who is too clear-eyed and truthful and aware of
my manoeuvring, says quietly, 'I'm going to need some time.'

And I let it be. We don't speak of it for the rest of the day.
I go outside and check where in the garden it can been seen
from. Not in the top beds in the front, not in lower ones,
the roof of the house blocks it, mostly, until you stand at
the bottom of the garden and look up. You don't see it until
you cross to the glasshouse, or go up the gravel path into
the Wedding Field. Then, it sits there in your eyeline, and
does need some accommodating. We know we have to get
our minds around this. The turbine won't be going anywhere
(*except around*), so now we have to fit our minds in and about
the reality of it, somehow. In 180-degree vista, it towers over

everything else and draws your eye, even though it is still only that engineered phallus, and is not yet moving. I stay out by the glasshouse and look at it. The first phrase that comes to mind is the one from *The Handmaid's Tale*, 'Under his eye'. And that's what it feels like. This sense of something now sited, up there, like an overlord, on the hilltop, looking down over this farmhouse and garden that must appear like relics of an earlier century, as I suppose they are. This is the first totem of the twenty-first century in this landscape, and that too will take some accommodating.

In this of course, we are out of step with many. You can now see nearly a dozen turbines on the horizon as you drive out from the village, and in the post office today a man twenty years younger than me says, 'I love the look of them. There's a real Star Wars-y feel when you see them all marching along the sky, isn't there?'

Yes, I don't say.

At midnight, a few nights later, under powerful floodlights that illumine everywhere around, the blades are attached. At forty metres long, each blade is almost half the length of the tower, so it elongates the presence, and now, in a small wind, starts turning. When I walk some friends out to their car after dinner, we see it, lit and enormous, with what appears to be a drone circling it. But, when I come in, I don't say anything to Chris. Tomorrow is time enough.

When I lie down in the dark in the bedroom upstairs that turning image is with me. The thought of the giant thing, 500 metres away, slowly rotating, like the clock of my own life running on, is oppressive. It takes a good while to come to a counter-thought: *Be more in the now. Live now.*

*

But that is more easily said than done, and a morning comes when you wake and you know that you've lost heart. Just that. You've lost heart and the battle you've been fighting for decades, to live a certain kind of life, you realise is a hopelessly, helplessly losing one. The realisation takes your energy, hits you in the pit of your stomach with What-have-you-been-thinking? Because you are falling into that place of profound aloneness in which the green quiet you live in speaks only of that, and the wonder and beauty are not available, and the only thing you can do is to get outside and let it be that way, don't fight or dispute it, but *do something*, however small, deadheading, say, or weeding, and if you give yourself to that this sadness will ease and yes, pass.

*

For a week the second turbine can't go up.

It is too windy.

*

Today, I am aware that I risk being that hackneyed figure, a gardener who says, 'My God, what a delicious thing is your own tomato!' But aware too that I have grown old enough now to say, *Go ahead and be the cliché, what of it, that tomato is good!*

And here, in October, they are more than plentiful this year. The glasshouse and tunnel are full of them. The orange cherry ones are astonishingly flavoursome. When you slip one in your mouth, the sheer sensuality demands all your attention. You are, by force of nature, brought into another life, or the life of another living thing, so separate from your own, so other, and yet so contingent on your care, that it is in microcosm the essence of gardening. Or so it seems to me.

This is not the same thing as when you stop at the farmers' market and try something off the displayed produce – which may well be superb, (all of Jason's are) and which you can fully enjoy – for it can never give you the other part of the pleasure, which is in fact yourself, and your being part of the world. The part that says: *You grew this, in this is your time, your love.* And all the invisible ties, elemental, primordial and human, that are bound by the simple fact of that.

Well, not to put too fine a point on it, but I believe it was I who 'grew' them and minded them and pinched them and sprayed them with biological fungus control when needed and fed them and guided them up the scaffold that I made of green twine and bamboo canes, nearly toppling over several times. (A thing I won't tell the bone specialist.)

   But Himself did bring in the soil, and the manure, and carry the rainwater from the butts each day, so maybe it is more accurate to say 'That tomato was us'.

All of which is just to say: That tomato was *good*. (My imagination, of course, always goes too far, thinks nothing is ever sufficient to convey rapture and wonder, and I am not far from 'Let me honour that tomato with its own eulogy, which will – already has – outlived it. Let me bring back the deliciousness of that particular tomato in some sentences here, which no matter how grand or grandiose will not, could not, do it justice. Because you simply cannot *imagine* how good that tomato was; its goodness was understood first in your senses, in the feel of it sitting in your palm, the flawless smoothness of the skin, the curve, (*the curve!*) the redness, (*what sunlight it had captured to become so red!*) its give as you bit into it, the juice rushing to the corners of your lips, the flavour already crossing your palate, already shooting into your brain and translating taste to thought: *This tomato is just excellent.*')

There is also the idea of a thing come to perfect ripeness. A promise fulfilled. How many times through July and August did Chris and I ask, (as Monty Don had wondered on the BBC) 'Is this going to be a bad year for tomatoes?' They had seemed extraordinarily slow to get going, and then poorly leafed for so long, some furling like dried things, then slow to flower. All manner of reasons were suggested, the soil, the sun, the seed, of course the watering, as well some un-reasons, the stress in their gardeners, the uncertainty, the fractious air the garden itself must have felt this summer with the diggers and the trucks and all the rest. Some of what followed, the close-looking, the silent assessing, the consideration of why plants were not prospering, is part of every gardener's experience, and though you might have grown something for thirty years, last year's tomatoes count for nothing. Nothing in nature is guaranteed, the 'bad year' a universal diagnosis and from long experience vouchsafed its place in the gardener's lexicon.

'A bad year for the tomatoes' is not the end of the world. No sense of encroaching calamity. All artists live with a history of failure and disappointment, as does anyone trying to grow a garden in the west of Ireland. The point I am trying to get at is the fact, and mystery, of ripening.

Because, despite a rain-sodden August, and rain-doused September, the tomatoes did ripen. From somewhere we were not generally aware of, they stole enough sunlight. (It's a thing I like to think of, their relentless hungering for and harvesting of light and heat, how out of each day the plant silently seeks what it needs to feed the fruits, and how skilled it is at finding it.)

Tomato teach me that!

And gathered sunlight is what you taste in a ripened tomato. That we now bring inside a bowlful every day is doubly satisfying – satisfying is not the word, enriching is closer – in these darkening, cooling days of October, you have the sense of the summer on your plate, and palate, and also that deeper thing, that feeling of covenant, of promise kept, that after all Chris's hours out there, the plants delivered, and there is, well, *this*.

From New York, our daughter sends a message: Stop using those teabags.

She attaches an article from the UK's Independent *that says 96 per cent of teabags contain polypropylene. So, when you drink the tea and when you throw them into your compost bin you are adding microplastics. The article highlights the difference between packaging that is listed as biodegradable, which means it will break down, but only under the right conditions and mostly at an industrial level where high heat is available – so not in your compost bin or your garden – and compostable, which is safe and will decompose naturally. N says he's old enough to remember the controversy when teabags were first introduced in Ireland, when they were derided as 'dust', not 'real tea', and even 'plastic tea', and the ad campaigns that aimed to convince people teabags were not only easier but cleaner. The tea companies were doing us a favour. Reading this article, we both have the sense that secretly we must have suspected the truth about the polypropylene, but simply for convenience, carried blithely on.*

No more. When your daughter talks, you listen...
Little by little, the young may yet save the world.

Sometime in our first year here, I'm guessing at the start of autumn when the dark evenings came, Chris and I went to Dublin in our ancient Peugeot and brought back a deep and squat old television set that my parents were throwing out. It seemed anachronistic, set up alongside the ten-foot hearth with the grate on the floor, down-puffs of turf smoke sometime obscuring the screen, and there was some discussion as to whether it was a betrayal of sorts. I can remember the undisguised surprise, not to say dismay, when American visitors arrived and exclaimed, 'You have a *television*?' It didn't matter that it only received two channels – and only after J J Keane had arrived with a length of copper wire and a piece of timber that he stuck in the upper rafters as our 'aerial' – there was a sense that we were supposed to have turned our back on all that. What visitors wanted, I think, was for us to be living in a kind of designer poverty, or inside a house in a folk park. I understand why, and between Chris and I there were certainly discussions as to how we should be living in this place, and in such a way that was true both to ourselves and to it. That is a negotiation that never actually ends.

Well, anyway, we got the television and set it up, and in the dark evenings watched the hazy pictures that travelled from the rafters down the copper wire. The news broadcast then was not what it was to become; it was not entertainment, it was terse and brief and, literally, matter-of-fact in a time when fact held supremacy. So too, the weather forecast. This was done always by a man always in a suit with a pointer and four maps, each about the size of the forecaster's head. On these maps the country was smaller than the weather, which mostly consisted of a number of stuck-on tabs that looked like speech-bubbles with arrows or exclamation points in them and made it look like we were under aerial attack, which I suppose we were. When a friend, who was a CBS producer, visited from America, and saw the weather forecast, he burst out laughing. With his pointer, the forecaster kept anxiously tapping the various arrows of approaching rain, until one of them fell off.

At that time, from an honourable history of profound fallibility, people here didn't attach too much belief to the forecast. They watched all right, but more out of curiosity, the way you might something going wrong before your eyes. Nor did anyone blame the forecasters in the way that became customary later when the technology promised fact and the world became a little harsher. Then, in these parts at least, I think it was understood that we were an island in the Atlantic, that nature was contrary and greater than science. So, if we were up in Dooleys' house, say, for a visit – Michael and Breda then in their seventies, the kettle sitting on the edge of the ashes in the fireplace and a pale griddle bread with charred intricacies being cut – there would be some talk of the weather, but notably, never with complaint. Now, some of this was generational – the older people had lived through enough to have deep layers of acceptance in their skin – and some of it probably bound up with religion, forbearance, stoicism and a creed of not questioning the Almighty. Also, people like Michael and Breda had lived through a time without any forecast other than the birds' and had a folk memory of their grandparents speaking of The Night of the Big Wind, which had taken thatched roofs and thrown cattle into ditches in a storm of biblical force that must have felt like the end of the world.

And so now, whenever we hear of a big storm coming, it is of those older people that I think, and the storms of long ago that came unheralded out of the vastness of the Atlantic and were on top of them before they knew it. The hectic and havoc of that, cattle to be housed, hens to be got in, a world of things to be tied down. What terror there must have been in the unknown. What rosaries were said in the candlelight as the winds howled.

Well, now of course, we know everything. And about a week before he was due to arrive, we began to hear about Lorenzo. We were visiting our friends, Jack and Colette, over in Ballyea when Jack said he was keeping an eye on this one hurricane down in the Azores. 'Lorenzo. Coming our way, I think.'

And sure enough, on the green screens of all the weather forecasts over the next few days, there he was, Hurricane Lorenzo. He'd make landfall on the west of Ireland.

I'm in two minds about the naming of storms. On the one hand, it seems a further gesture to entertainment, the storms have become 'characters', as if human-dimensioned or -natured, which has the effect of diminishing or containing them, taking some of the mystery, force and actual wildness of nature. Wild things do not have Christian names. On the other hand, maybe this makes the storms less frightening. Something in me doesn't like the naming, but I can't get to the bottom of why and concede it may be just me.

'If Lorenzo is as bad as his sister, Ophelia, last year, we're in trouble,' a man at the petrol pump in the village said the day before the storm was due. (Mercifully, I didn't point out that in fact it was Laertes who was Ophelia's brother.)

'Still four hundred kilometres away,' I said.

'He is. Moving fast though.'

We both looked at the sky, which was perfectly clear, and the hurricane seemed like a story.

'Better get ready.' He smiled.

And that's the thing, a twenty-first century behaviour, after the satellite tracking and the green-screen projections of an expressionist, harmless-looking swirl, the reporters standing in raingear on coasts where there is no wind or rain yet, the readying. All afternoon Chris and I do our usual and go about the garden checking canes, tying up the late-flowering plants like the Michaelmas daisies and the black-eyed Susans, taking indoors pots that could fly, and into the cabin putting the metal summer table and chairs. The afternoon is perfectly lovely, so still that you imagine the stillness is strange. Because half of your mind is out there in the ocean – where Lorenzo is – and that half tries to picture the actuality of that, the wild ferocity of the storm in the rising and falling walls of the grey ocean, you imagine that this stillness is a result of that, as though all the wind has been *sucked* west to join in what is happening out there.

I come in and check if it is still true that the hurricane is coming. It is. On the radio, while waiting for something to actually report, the talk is whether this is to be our future in Ireland, that warming seas will send us more violent ex-tropical storms. I fill buckets and pots with water, find the candles, the torches, all the things people would have done in this house a hundred, two hundred years ago. I think of it as fitting. I think no number of centuries of progress or technology can out-manoeuvre the reality of a vast Atlantic storm, and the closing down of your day, this reduction to just getting ready is right, and respectful.

But the truth is, no matter how many you have lived through, a storm cannot be grasped in any physical way until it arrives in your senses. Your brain understands the idea of it, you have flash images and reference points, and the knowledge that you and the house have lived through many – we have – but in the fore-stillness you can't feel it and it is still hard to credit. You ready in blind faith then.

Chris and I secure all we can, leave the garden beds filled with bamboo canes poking up, like splints or crutches, and come inside to wait. Our neighbours are all doing the same, and there is a sense of the community, like all those along the western seaboard, huddling down. Nothing comes or goes on the road that has been so busy during the construction next to us. The crane has been lowered out of the sky.

In most storms here, the electricity is first to go, and so as the evening comes on there's a candle lit on the dresser, waiting.

Just before dark, it sounds like the sea is in the pine trees across the road. The tops of them are swaying wildly. There's a kind of language in it, whispers and sighs soon becoming louder. A kind of roaring gathers. I listen for breaths in it, stops and restarts, pauses, like in a living thing, but the storm doesn't pause and the roar is one long continuous sound. The storm proper is just arriving. And now, suddenly, Chris puts on a wool hat, two coats and her boots.

'What are you doing?'

'Going outside.'

'You can't. It has started. It's dangerous.'

'I can't bear it. I can't bear being shut in. I have to go outside, or I'll go crazy.' She knows it's just as, if not significantly more, crazy to go outside just as Lorenzo has landed, but she's serious and in her face I can see that there's more than just the confinement of the storm, there's everything she's been battling. 'Just for a bit,' she says. Then adds, 'I want to go on my own.'

'Why?'

She considers that for a moment, and then, with the twisted logic of what I'll call storm-mind, answers: 'Well, that way Lorenzo will only kill one of us.'

I stand at the door to watch her. Into the huge wind now she goes, a small figure who heads down the garden and inspects each bed, something like a general in the first stages of battle.

*

'Those Brussels sprouts look like cabbages. Are you sure they are not cabbages?'

'They said Brussels sprouts.'

'I'm not sure the plants heard.'

*

In the middle of a still clear day, with a sharp crack, the sound of a gunshot.

It travels some distance, I don't know how far, but it arrests me in the garden. There is just the one shot, but in its wake, everything is quieter, as though tightened, and the thought that comes to me is: I can feel the shock of it into my bones, so what must it be to the living things that don't know what it is? The crack is not like the sundering one of a tree, I have heard that and know the harsh slightly elongated 'A' vowel of that, which is shocking, a sharp ache-cry, but the gunshot is something else, because of the blind swiftness of it and the reach of the violence.

I know the shot is from someone in the gun club. Always at this time of year there are a few out hunting pheasants. They come with small dogs and set off down the valley and into the fringes of the forestry. The dogs rise the pheasants from their squat hidings and the gunshots go off. It is part of the life-rhythm of this place, has been since forever, and I have come to accept it as such.

But still there's something that shudders and recoils at the sound. Never mind the arguments about living and letting live, I am speaking just about the instinctual response of your nature. Or mine anyway. The gunshot cracks open the quiet, and in its aftermath all of me is sort of waiting for a second shot. But this morning it never comes, and gradually the quiet heals over, and I carry on doing the autumn tidy-up.

It may be three hours later, bringing a barrow of sycamore leaves over to where I am trying to make a leaf mulch, my heart stops when, up from the long grass that is never mowed, there flies a pheasant.

There's nothing quite like it. There's nothing like the start in it, and in you. The sound it makes is a kind of throttle and thrum, a bird-engine that is partly the sound of a wound propeller on an elastic band, only more basso, because it seems to come not only from the wingbeat but from inside the deep breast of the bird. It can't be, but it sounds like a startled heart beating wildly. And the pheasant's flight is not like other birds, not easeful and instant, but rather an effortful frantic climbing that the bird seems unsure she can make.

She flies up and across a short distance in what is called the back haggard. I want to apologise for disturbing her in the dreamy grass. I want to say 'I have no gun' but she goes across the wall and into the meadow, and both of our hearts are not the better of the encounter for a bit.

But mine is enriched. By the beauty and the wildness of the thing. This is the wealth in this day, and I am grateful and illumined inside in the quiet ordinary solitary way. For the next hour or so, I am one-part endowed by the flight of the pheasant and one-part listening for the approach of the gun club.

Because pheasants and I have history here. For several years now, at this time of year, I have been tossing some stale bread over along the western wall. And the pheasants do come for it. I can see them from the kitchen window first thing in the morning, sometimes one, sometimes two. And I think I read somewhere that a pheasant, though wild, will remember everywhere that food was found, and what's more, what's too astonishing not to be believed, *that memory passes from one generation to the next*. It is the same way, I was told, bats return to the nesting holes of their great-grandfathers – or is it grandmothers? In any case, it passes into the blood knowledge, locking animal and place in a kind of mutuality that I find deeply heartening.

Enough to say that, in the aftermath of meeting the pheasant, part of my pleasure was in the thought that I had fed her grandfather.

Three vans of the gun club pass down the road later. They do not stop. The pheasant is low in the long grass of the meadow, one startled flight enough for her for today.

*

During the night, under massive floodlights that light up the country like an airfield, the second turbine goes up. When I come downstairs to make the tea, the grey tower is standing on the hill. Because of the perspective from our kitchen it seems three times the height of the first one. It towers over us. And it will be a third as high again once they attach the blades. It takes my breath away.

It is a low day.

But in the afternoon, a young neighbour, Noel Downes, knocks at the door with his five-year-old son, Ruaidh.

'I brought Chrissie a cherry tree,' Noel says. He doesn't refer to the turbines, just says, 'Thought it might lift the spirits, you know?'

It is such a generous and understanding gesture, it catches me unawares and blows my heart open. I think of Mary Breen and the quiet civility and generosity of the old people when we first came here, when Noel was a small boy. I can see her smiling, her arms folded, and hear her say, '*Now for you, Niall.*'

And by this moment, something I thought lost is restored.

'Do you like tomatoes?' I ask Ruaidh.

'I do.'

'Come get some so.'

*

About ten years ago, we ran the first Kiltumper Writing Workshop. It is not in my nature to believe I have any expertise in anything, knowing a more or less constant battle with doubt, and still trying to write one good book. So, I was slow to jump at the idea. But, as in all things, Chris led me towards it, first, because she said I had something to offer, second, because it was a way to bring us, *in our own place*, the company of writers, and third, as always, we needed the money. We did not advertise it widely, only listed it on the website, but for that first one, twelve writers came from a number of countries. There were only two I think from Ireland, an accidental fact that furthered the sense of the workshop in Kiltumper being somehow a guarded secret. It was not, but felt like it. The novelty of the thing only came home to me when I was down in the village and was shyly asked: '*Have The Writers arrived?*' Two hours later, our great friend Lucy sent her husband Larry up to the house with a freshly baked fruit cake. Larry, one time the school principal in the village, is eighty-five now, and the finest man I've known. To him and Lucy, Chris and I owe an enormous debt of gratitude for their support and friendship over the years. Leaving the car running, Larry came to the front door with the cake and delivered it with a single line that required no elaboration: 'For the writers.'

At that workshop, the writers landed in our kitchen on the Friday night for drinks, for Chris and I to thank them for coming and to put faces to the work they had sent ahead. Chris had been corresponding with them for months, organising accommodations to suit budgets and generally assuring them Kiltumper was an actual place. She knew them all better than I did, and without her of course the thing couldn't and wouldn't have happened at all. There were writers from the US, Canada, England, Scotland, Italy, Norway and the Arab Emirates, as well as Cork. The next morning, the writers assembled around the open fire in the sitting room and we began. It was intense. There were group exercises and individual sessions, much writing and talking done, while just the other side of the door, offstage left, Chris was trying to make lunches for twelve people – each day three different salads, two new soups from scratch, and an afternoon cake – without making a sound. It was a remarkable moment in the life of the house. It was one of those things we might have imagined in the innocent months before we came here so many years ago: 'Writers will come and see the garden and the house and we will make them welcome.'

In all writing workshops you are handling intimacies, and there is an air of vulnerability, both of which I acknowledged in myself. But somehow, over the course of the long weekend, around the big hearth in the ancient house, an extraordinary thing happened, not only a bonding, which is natural, but what I'll call a sense of communal vocation. Maybe because each writer had had to set out for Kiltumper from a distant elsewhere, had had to set aside the time and find the funding, had had to postpone or pause their life for a few days in order to devote themselves with all seriousness to fiction, there was already a charge loaded on the workshop. But the thing that also came about was to do with aloneness. Every writer is alone with their work, always. Every writer, if they are at all truthful, knows that their work fails the standard of their own imagination, but they must keep trying. That standard is not measured by publication or reviews or prizes. It is inside you; it rose up the first time you read something that stopped your day or entered your dreams, and kept on rising every time you read something that made you say *That's just perfect*. So, what happened that first weekend with twelve writers gathered in our sitting room was a shared sense of the private, silent, struggle with language, the impossibility of success, and the acknowledgement that we were all alike bound to that same struggle.

Or something like that.

We finished late on the Monday afternoon. Chris had joined us, she sat on the back step, on her face a dazed look with the realisation that somehow, she too had pulled it off. With borrowed chairs and stools, mismatched plates, a prima donna of a water pump that had decided to act up under so much use, somehow we had pulled it off. She had said we could do it, and now here we were, in the last moments. When I looked at her, I could feel the deep part of myself welling. In our life, she has always been the daring, the visionary one. To the writers I said a last few words, thanking everyone for coming, wishing them well as they returned to their lone battles in different corners of the world. I said all I could say other than The End.

And there was silence.

Nobody moved.

The writers were arranged on two couches, armchairs, kitchen chairs, in a semicircle around the fire. I was sitting with my back to it. I closed up my folder. And still, nobody moved. Nobody looked at each other, they just stayed perfectly still. It was as close as I have come to a sense of spell. I didn't want to fracture it, nor could I stay in the seat any longer. I had nothing more to give, but knew that if I started talking again the afternoon would go on into the evening and the night. (And for a moment I could picture that, The Workshop That Never Ended, the writers that never left, they came to Kiltumper, and then…) I looked across at Chris, and, as always, she rescued me. She stood up and opened the door out into the kitchen.

Silently, I got up and joined her, stepping through the scattered written pages on the sloping floor in front of all the writers. We went into the kitchen, and stood there waiting for the spell to lift.

And then it didn't.

None of the writers came out after us. They stayed where they were, gathered around the big fire in the open hearth. It was as though they wanted to preserve not only the moment but the entire weekend, and knew that the moment they left our fireplace ordinary time would be switched back on, they would go their separate ways, lose this sense of solidarity that had built up over the four days and be once more in the lone battle of all writers, against doubt, against the eternal unanswerable question: *Is this any good?*

And so, without saying so, in a kind of communal accord, none of them got up to go.

In the kitchen, just the other side of the door, Chris and I listened and waited. She raised her two palms in a mime of '*What's happening?*' I didn't know what to answer.

'You have to go back in,' she whispered.

'No. Give them another minute. They'll come out.'

We did, they didn't.

When that first play I have spoken of was in rehearsal at the Abbey Theatre, the doorman at the theatre had told me he could always tell how successful or not a play was by the speed at which the audience left the auditorium and pushed open the doors to get into Middle Abbey Street. There was an exact mathematical relation, he maintained: the faster the leaving the worse the review. 'Feet don't lie,' he told me, with the peculiarly flat certainty of an older Dubliner. Well, on the opening night of *The Murphy Initiative*, there wasn't exactly a stampede, but we knew. I can still picture my parents in the lobby as the audience flowed out past them, and the sense I had of a public shame.

Well, here was the very opposite of a stampede. The curtain had come down, but no one was getting up from their seat.

'Go on, go back in.'

Going back into the room was one of the strangest feelings I've had in this house; it was the only time I felt like an intruder in our home. I really didn't want to go back in, because I too had the same sense, I knew something special had transpired over the few days, which was in essence the defeat of loneliness, the particular kind that all writers know. And though we knew the victory would be short-lived, everyone wanted to preserve the life of it a little longer, and I was no different. I was aware of the remarkableness of the workshop, not just from the point of view of our having pulled it off, of all these writers trusting we had something to give them, for their stepping out of their lives to come to a two-hundred-year-old farmhouse in a remote place in the west of Clare, but also in what the writers and their work had taught me, what they had given me, not least their part in vanquishing my own aloneness. As I have said, I was also aware that over thirty years ago, while in the glass canyons of New York, we had dreamt some version of this, and was not a little dazed with the realisation that it had now just happened.

To announce myself, I made a bit of business of opening the door, and stepping down the two stone steps into the room.

I think there was a moment when it must have seemed I was going to resume the workshop, and that time-outside-of-time, Kiltumper-time, would continue.

'Would anyone like a cup of tea?' I didn't say *before you go*. But the mundane will always shatter the sublime, and at last the group stirred. They came into the kitchen and continued to talk to each other, animated and intense, in a way that made it seem they had known each other a lifetime. None of them were particularly interested in speaking to Chris or me, but the kitchen was buzzing with their talk. There was an after-theatre sense, as when a number of people have privately shared the same experience. I realised something extra, for what they had shared was perhaps the most personal, intimate parts of themselves, the thing that came from their deepest selves, their writing. And so, for these hours at least, they were bonded profoundly. All of this, I now understood, I had underestimated, or not estimated at all. We had some responsibility here, so as the tea finished – and no one made a move to leave – and someone offered to make more tea – 'I'll make a pot, will I?' – and it became clear that the workshop *could not in fact end*, Chris stepped forward and said, 'Well, we are exhausted, so we are going to go to Doonbeg for dinner.'

Briefly I thought someone would say 'Fine, go ahead, we'll stay here.' But Chris countered this by quickly adding, 'Would anyone like to come?'

It was a masterstroke, and maybe the only way to get everyone to leave the house.

And so, the workshop transferred itself holus-bolus to Doonbeg. There, after a long dinner, it was Chris and I who managed to leave. We thanked and embraced everyone then slipped away out into the wild Atlantic wind, leaving the workshop to find a natural way to end.

Well, in the ten years since then we have tried to run two writing workshops here every year, usually one the first weekend of May, and one the last weekend of October. In that time more than a hundred writers have come to Kiltumper. A good number have had novels or stories published since, or gone on to MA courses in creative writing, but more importantly I think, they have felt some encouragement and gained some insight. And certainly, part of what they experienced has been to do with this place itself. It is deeply satisfying and feels, well, in that well-worn term, organic. Something that has grown naturally here. I am proud of Chris for making it all happen, of the house and garden for accommodating it, and in quieter moments filled with gratitude to all those who have come.

This October then, in response to some requests, we held a 'masterclass', for ten writers who had been here at a workshop before. One came from Ireland, the rest from England, America, Germany, as well as Lisa, who came from Norway, and who had been here for the very first workshop all those years ago and since published a first novel. It was a weekend of blown rain, the garden going darkly into November, and the road outside full of potholes and loose gravel and mud after all the turbine traffic, and now a line of orange cones where the road had been widened and the ditches removed. I worried what those returning would think of the place, which may have retained a glow in memory, and of the two writers and rain-gardeners now grown so much older.

But again, I underestimated. For the weekend took on an air of affirmation. We had all carried on. That was the point. We had not given up. Despite all of our varied and multiple failures, we were all still trying to write a better book, and Chris and I were still trying to have people come to our home and share whatever we knew.

'It's even better than I have remembered it these ten years,' Lisa said. 'But one thing,' she smiled at the memory, 'this time I am leaving *early* on Monday afternoon. I am taking Kiltumper with me.'

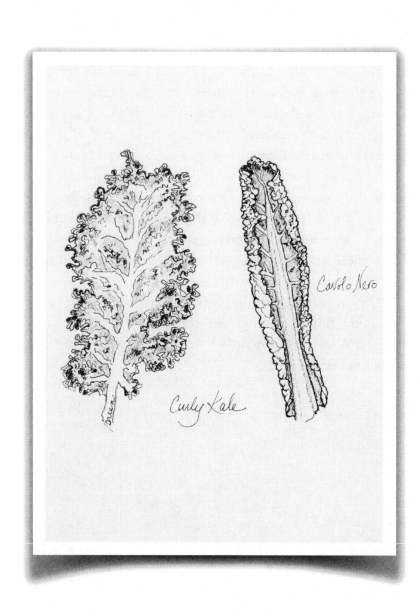

Curly Kale

Cavolo Nero

II

November

And there comes a morning when you and the winter are looking at each other. You look with all of your years and all of your senses. You know it is coming now and somehow, by an act of kindness inbuilt in creation, you may have even forgotten about last winter, your bones have forgotten and your blood has forgotten, and all through May June July August September you have forgotten. Like every other plant and living thing in the garden you've been too busy with the heady stuff of growth and bloom and even going-off-bloom (which has its own headiness and dance). But one morning, here it comes again. Winter comes not with the dramatic record lows that are the common-stance of our times, not with an early snow like in New England or an ice wind, but first just with an elemental darkness that cuts off the afternoon at the knees so it genuflects to the sun setting. Still ahead, but now remembered in your toes inside your boots (*I must double my socks*) is the iron cold of the earth itself, that, because here it has an ally in dampness, can bring a cold into your bones that cannot be measured by thermometers. I've already said, but will say again, the older I get the more

I appreciate that winter is not a thing to be taken lightly. I know this may seem absurd in a man of sixty – which is another man or woman's young – but still, I think without thinking, instinctively, today: *Here comes the winter. Who or what will die before it is done?*

But, soon enough, the garden itself undoes the furrow of my brow. Walking around it with Chris at four o'clock this afternoon, she lists the plants that must be moved or divided over the winter. There is in all gardens this sense that next year's expression will be an improvement, because the garden itself has taught us, we have learned that little bit more. There's an inbuilt optimism, and more than anything the sense that we live within a round, the world's journey cyclical, and the winter a grace, a pause-time for redrafting and putting right all that we have got wrong. That it will be ever so, that there will never be a time when we look and say, 'That's it, the garden is *done*, nothing needs to be improved for next year,' is part of that grace too, for it carries us forward, dreaming again, hoping again in all our time on the planet.

And this afternoon, the more we plan the more the garden says: *Fear not. Come April, we will be laughing with buds again.*

❧

When the time comes to go inside, darkness falling, we harvest lettuce and a large bowl of tomatoes from the glasshouse; the yellow cherry ones are definitely the best this year. Nobody can eat one without saying so. Tomatoes and lettuce in November, in Kiltumper, these are triumphs. Small ones maybe, but still triumphs in my book. My harvesting skills are improving. To have the glasshouse and polytunnel still producing to the last days of the year makes me happy.

Sometimes too, in this peculiar, particular kind of life we are living here, there is a day like this: at seven o'clock in the morning an email arrives to say *This is Happiness* has been chosen as one of the books of the year by someone in New Zealand. It is a long-distance thought, but a lovely one. At eleven, from America, a cousin of Chris's sends a photograph of a page in *Real Simple* magazine showing their five best books of the year – *Happiness* one of them.

To think of the words written here in this farmhouse now in those distant places on either side of the world makes me feel a deep humility and gratitude, a kind of writer's blessing. But there's no time to consider this any deeper. By this stage, Chris and I are out planting a copper beech hedge and twenty alder trees with the help of Tom and Paul, Chris's co-workers at the garden she maintains on-and-off at the grand old hotel in Ennis, The Old Ground Hotel. These trees and hedge are to be part of the defence strategy against the turbines. The hedge will go east-to-west along the bank outside the kitchen window. In years to come it will be what you first see when you drive in; more importantly perhaps, it will take your gaze to the line of it when you look out from the blue couch in the kitchen. It will never be tall enough to hide the turbines beyond, but the hope is it will give you something closer and more beautiful to see first. It will also make us feel a little screened, although at night the flickering red lights will be above the tallest hedge and tree. We plant the alders out along the boggy edge of the Big Meadow at the back of the house. Each hole we dig out gleams with wet muck, watery seams silver into it, but Chris spies worms in the hole, exclaims for them in fact, and bends down to make sure I don't spade them in half. On her knees alongside me, she holds a pair of them waiting for the hole to be dug.

'Really?'

'Yes.'

'OK.'

Despite their honoured place in Irish mythology and folklore – the alder a fairy-tree, protected by water spirits, a symbol of courage and spirit, in mythology the first man made from an alder – generally, hereabouts the alder is not considered very noble. I have heard them cursed as weeds for their easy self-propagation, and appearing in places where they are not wanted; but they are wanted here. They are fast-growing, and prosper in the damp ground, and we are planting them in the sightline of where the turbines can be seen from the glasshouse. They too will never be tall enough to hide them, and, at five feet today, we know we are planting them for whoever comes after us.

⤙⤚

'But will *anyone come after us?'*

It's something I think about a lot now. Planting the trees for whoever will, or will not, come after us has sadness and loneliness in it. We are not different to other ageing couples all along the west of Ireland – the Wild Atlantic Way – who, for countless generations, have seen their children emigrate (like the eleven children from this cottage in the 1900s). Our first left after the 2008 crash, fresh from a four-year degree from art college, because of no prospect of a job to go to, and our second a few years later with a law degree. It's part of the nature of this rural, empty place, and although there are exceptions (one family in our townland have all but one of their seven children living back in Ireland, having ventured as far away as Vancouver and Melbourne, and another family have all four back in the country) there is some hope that if broadband ever comes — ever — it will make a difference. I'm only kidding myself. It'll take a lot more than broadband.

Our house is so old, with walls made of stones. The floor in the old part of the house has no real barrier against the cold earth. Where we can, we have laid wooden floors. Our windows are for the most part single-glazed. The new part of the house has insulation in the attic and that helps our upstairs bedroom stay warm when the heat from the stove rises. (We have a beautiful kitchen, now. So there's that.) Needless to say we spend a fortune trying to heat the house, so during winter we tend to stay by the stove at night and close the five doors that lead from it into other places. When the sun shines, the conservatory is as warm as a summer's day.

In my heart I am hoping one day things will come full circle and my children and their husbands will return to Kiltumper. Or at least to Ireland.

And then we can all plant more trees, together.

Maybe one day.

Planting trees feels like the right thing to do *for the place.* An argument that is neither straightforward nor brief, and Chris and I do not make it, we just go ahead and spend the morning planting the trees, the two horses in Lower Tumper standing watching and any number of birds waiting to see if those worms went back in the hole.

By early afternoon the trees are in and staked. We thank Tom and Paul and have a mug of tea together and some of Lucy's curranty cake out by the front garden. It's undeniable there is some virtuous feeling after planting trees, and I am not cynical enough to deride it. It feels good. So be it.

Then I am hurrying inside for a quick shower, for a scheduled phone call from London with Polly, the woman who is hoping to direct the film of *Four Letters of Love*, which, like all independent films, may or may not ever happen. But, importantly, you act as if it will, because in this business maybe dreaming makes happen. Garden clothes in a brown heap inside the back door, mud-coated wellingtons outside it, I see an email sent to Chris and I from Jo, to say the novel she workshopped here eighteen months ago has just been shortlisted for the Costa First Novel Award and she wants to thank both of us. It's marvellous. A small, private, but nonetheless potent landmark in the organic life of the place. The film phone call lasts an hour and I am inspired, in the original sense of in-spirited, the something that enters you and you believe again in the possibility: *This could be wonderful.*

I hang up and go outside. Chris of course has stayed out working. She has found a dozen different things that needed doing, and in the process of doing them finds a half a dozen more.

The afternoon light is dying very beautifully. I think if I was given only one hour on earth, I might choose this one. It is a threshold time, almost between worlds it seems to me, and where I feel at home. Those moments when the light caught in the trees is full of gold, the only time I think of the word *burnished*, a light you can't believe exists, is actual, even as you are witnessing it, a light that cannot be captured by your camera despite Chris's numerous attempts to catch and send it to the children in New York, a light that demands you look at it even as it is quickly going away. The bare limbs of the ancient sycamores turn black as Japanese ink drawings against the sky. Up in Lower Tumper the standing horses go from chestnut to grey, their forms becoming fugitive and soon fusing into the field.

Over at the tap-on-the-tree, I hose the deep mud off the sides and soles of the wellingtons – 'Save the mud! Wash it into the bed' – and then the spades, the handles of which are too ingrained ever to be clean. Together, in the new dark that is still thin enough to see, we do a quick tour of our new hedge and trees. Then we have to go inside and get ready, because this evening the twelve members of the book club that has been gathering in front of our hearth once a month for the past fourteen years will be landing, and nothing is ready. There will be a warm company, some lively discussion and more laughter, as well as wine and tea and cakes, and after, when Chris and I finish tidying up around midnight, she tells me the *Washington Post* has just chosen *Happiness* as one of their books of the year.

And that, as they say, was that day.

～⚬✦

As N has said, we have been walking this road outside our house since the very beginning. With our son and daughter. With visiting family and friends, and now just together. Our children learned how to run and how to cycle along it.

With the onslaught of the turbines in mid-west Clare – it seems everywhere you look – I don't believe our grown children will ever come back, except for a holiday. Then they will walk this road with us and remember what once it was like.

A few colourful leaves still hang on as I walk out, passing the land to the north, but looking away to the south. I'm almost getting used to the turbines standing like silent sculptures and some days I can say they are pretty, although, once turned on, their sound will not be pretty. The red berries of the whitethorn are like a thousand beads brocaded onto the thick coat of black branches and I closely look at them. The sloes along the hedgerow are tempting as blueberries, sprouting along the razor-straight twigs of the blackthorn. Fuchsia still blooms, somehow, and dangles like a hundred red and purple jewels, which seems impossible in November. And underneath the damp and shady hedgerow the hart's-tongue fern (Asplenium scolopendrium*) grows. There's one growing in the wall of the opening from the old cabin into the front garden.*

November comes to a wet and windy close with the first hailstones battering on the skylights and on the conservatory, bouncing on the patio and ricocheting across to the lawn like white confetti. I wonder if it is true that a wealth of sloe berries and haw berries on the thorny bushes along the Kiltumper road means a long winter is ahead. At least the birds will have plenty to eat.

I try to see beyond the changed landscape and into the cycle of the seasons and life and what is required and will be required of me as the landscape continues to change and as I continue to change. I won't be able to garden the way I have been. (I have in some ways overworked myself. No, in many ways.) What will become of the garden then? It is the thing that has kept me here, so far away from family when I was in my thirties and forties and now with both my children gone to NY, it gives me a connection that is otherwise missing. Now that the turbines have hijacked the countryside like whirling dervishes I fear County Clare is becoming a place of just forests and wind farms with people dwarfed in between.

Country living teaches you things. It teaches you about darkness and stars, about sunlight and silence, about things out of your control like when the wind uproots a tree, it teaches you that the berries and the birds are wintering with you and perhaps wondering like you what is happening. It teaches you that the road will always be there, ahead of you and behind you.

Going out for early-morning turf – it is suddenly very cold, too early to be this cold, and the fire must be lit first thing now – I startle five young pheasants and their mother. The mother cries out and takes off. Whether she is the same one I met at the haggard last month I cannot say, our acquaintance only in heart-racing. The young ones rise about four feet off the ground and sail across the road, one of the young hitting the top bar of the Downes' gate across the way.

There was a thud and then silence.

I go across to check, and watch the young one scutter-fly down the slope and into the long grass, where I leave it to recover its heartrate from what Yeats would call the fanatic.

This evening, as the light dies before five o'clock, a young fox comes up the garden to the glass of the door. Light makes its eyes dull orange as it looks in to where I am standing perfectly still, looking out, both of us likely thinking of the young pheasants.

Truly a surprise, how cold it is. This is the cold I associate with January or February, when it is tempered by being able to count the days to spring. For several days there has been a wind from the east that would skin you, a phrase that renews itself in my mind, the sharp edge of the wind coming now to pare away at my ears. This cold requires your attention. But once you have given it that, hatted, coated, hands pocketed, you can appreciate the polished quality of the light that has come with it. These are gleam-days, no clouds, sky blue as the Virgin Mary's stole, and the grass silvered in the dawn. It is also stoppingly still; no matter what you are doing, in the garden the stillness itself stops you. It feels visited on the place, or as though a veil has been lifted. You can see more clearly. I know that sounds absurd. But the stillness in the November garden is first about seeing. In this shone-in low light, the sun white not golden just above the forestry, there is an architectural elegance in plants already well past bloom. Chris delays cutting things to the ground until into the new year, and so the beds are filled with stalks and seed heads of different browns, ochres and blacks. And in the stillness today, strikingly beautiful, some of which comes from the knowledge of the fulfilment of a cycle.

Standing at the front of the house – trying to be perfectly still in the cold for ten minutes, not easy – I just look at the old plants, their lineaments lent a spectral grace in this astonishing light, and I think: *Take note of this moment.* That's all. For it is everything.

*

This evening, Thanksgiving, turkey in the oven and some friends coming for dinner, there appears up on the kitchen windowsill a large cat. It is the cat from back the road, left to its own devices when the elderly woman who lived there went into the nursing home in town. We have passed it on the road often, and sometimes tried to see if it wanted to come back with us. It never did. Until this evening. Whether the cat smelled the turkey, or got tired of being on its own, I can't say. But we wanted a cat since our last one, Tiro (Cicero's servant and transcriber, Cicero the cat we had before Tiro, too many over the years to catalogue them all here) died. So, tentatively, I welcomed the visitor at the back door, and, tentatively, the welcome was not refused.

'If he stays, we'll call him "Thanks",' Chris said.

⌒

N is planting garlic in the cool ground today. Thanks, our adopted cat, and I, watch him from the gravel path. Thanks seems sceptical. A foolish endeavour, he meows, unlikely that anything will grow in this weather. 'I know,' I say to him. 'But it's an experiment. Let him try.'

A cat is good company in the garden. We have always had one, and I am hoping Thanks continues to stay. He seems to like N, and N likes him, a first for Kiltumper. For I have been the cat person here. Now N is a gardener and a cat person! What next? Oh, and he's turning into a good cook.

Thanks continues to watch from the gravel. Behind him the vast shapes of the turbines on the hill are not yet turning, but they are there. Will I ever get used to them? I can't imagine that day. Never mind, for this moment, I can't hear them. Instead, looking like an unkempt mad gardener in two coats and what N calls my Russian hat, I concentrate downwards, on the ground, and fork out dandelions with my Japanese knife.

⌒

Hares are dying out. Some small discussion of this this evening. There are some hares still seen from time to time on the Frure road as you leave the parish, but nothing like the presence they had when we first came here. Then, any walk back the Kiltumper road might bring you face-to-face with one or more. Your heart remembers the jolt it got, and your mind the speed as the hare vanished, leaving only the memory of itself in the grass. Because of their beauty, because of their agility and bound, because everyone admires those who live by athletic grace, but also because of their longstanding place in Irish mythology where notably the hare can pass between this world and the Otherworld, and, in the story of Oisín, change into a beautiful woman, the absence of hares is saddening in a way that is hard to fathom, by which I mean in a measure unfathomable.

*

The quality of mist that surrounds the house and garden this November evening: grainy, both grey and rainy, textural too, reminding you air is a stuff, million-particled, made. A kind of comfort in that, though I haven't the clarity of mind to be able to unpick it.

*

Giving – dare I admit it – my wool hat its annual wash. The hat was hand-knit for me more than forty years ago now, and by the declension whereby good clothes eventually become farm clothes, a declension I think entirely natural and proper, the hat has become my winter gardening one. I don't treat it with any preciousness, and often, when my head heats up over some digging or dragging or the like, I sit it atop a fork or spade or hang it off a branch. It's a deep green, was knit in Paris, and hanging out there seems entirely at home and foreign at the same time, maybe not unlike myself at this stage. Because it accompanies me so frequently, and because I never think of it as 'dirty' I suppose, it never quite makes it into the wash with the rest of the gardening clothes. But today I plucked it off the place where it waits to go outside, and dropped it in a basin of warm water.

What emerged was a greenish sludge.

But, *all right*, I thought, that is the sum of what has been on top of my head this past year. That is all those days in this green place, that's all the endeavour and the dreaming.

Looks about right.

*

Today, the Minister for the Environment announced the government's Renewable Electricity Support Scheme, essentially an auction with state aid approval for wind farms and other green energy projects. In the last four years installed wind capacity in the country has grown by 50 per cent, the goal is in the next three years for it to grow by a further 30 per cent.

There is a notably Green, virtuous air about the announcement. The country is to have many more wind turbines.

Nobody asks the Minister where they are all supposed to go.

Because the answer is a given. They will go in what are perceived as the empty places of the countryside. Like Kiltumper. Some places of high 'scenic value' will be spared the turbines, the rest will be available to developers. That the effect of this will be essentially to ward off parts of the country as un-turbined, more or less protected parks, and change forever the nature of the rest of the countryside, with locked gates, new straight turbine-delivery roads is so upsetting to us that we have to turn off the radio mid-Minister.

The two turbines are standing high against the sky just behind us, like things from *The War of the Worlds*. They have not been switched on yet, so they only turn slowly, and we cannot hear them. But in their turning is an inevitable sense of clockwork, and the countryside steadily turning, away from the times we have known.

*

Another saddening note: this evening, on a visit to neighbours, the news that no new houses in Clare can have a chimney.

To us, who have known a ten-foot hearth and a fire lit each day for years, who in fact *have had a kind of relationship with the fireplace*, knowing some history of those born, bathed and fed alongside it, had Santa come down it (and had our children shout Christmas morning 'Thank you Santa!' up it), this news feels, in every way, disheartening. There will be no hearths in Clare, and I presume nowhere else in the country. In some profound way that I can't pin down, but is to do with the history of mankind on the planet, and the first making of fire, it feels tonight like a twenty-first century landmark, and a sad one.

*

At different times throughout this year, while waiting for my novel to be published, and received, in my mind there has been an ongoing battle between faith and doubt about the place of fiction – a battle really of my own place – in the world. *What is the point of all this hardship I have brought us to?* And *Is this what I should be doing?* are never far questions, as well as those about the place of story in a world where reality has become entertainment and Breaking News breaking every fifteen minutes. I am, by nature, I think, a storyteller. *By nature*, the important bit, I have come to understand this year. (I feel most 'natural' when I am inside a story writing it, most fully myself – although that appears a contradiction, it's not.) I remember William Faulkner dismissing some academic question about literature by saying 'Oh I'm not a literary man, just a storyteller.' And I have some sense of that, while also feeling at times that the kind of storytelling I am interested in, just the human heart in conflict with itself – Faulkner again – is verging on the antique. For me, a tell-tale sign that storytelling has lost some of its standing in the world is when an interviewer asks, 'How is this relevant today?' Relevance too low a measure to apply to any art, it seems to me. But I am aware, as always, that in this I am out of step, and usually make up some answer.

There is always an amount of inner to and fro while waiting for a book to come out. Because of the tightrope (which is made of a shoestring) kind of life we are on here, what we are actually always waiting for is to see if in fact we are going to able to carry on this way of living and writing for another while.

Well, this morning, late November, after a warmer reception than any book of mine has had in twenty years, I realised that some of that battle has been won.

For now, at least.

And that's good enough for me.

I took my tea outside and sat, hatted and doubled-coated, looking down at the frozen garden. What appeared first as absolute stillness was soon discovered not, and before long I was aware the stirrings of a new novel were just now announcing themselves.

12

December

It strikes me today that the part of my life I have considered the most ordinary – living here with Chris in this green hush – is in fact the most extraordinary.

*

What is the countryside for? There it is again, that question that has kept coming back to me this year, the year of the turbines. In my notebook I have written Robert Macfarlane's quote from *Landmarks*: 'We find it hard to imagine nature outside a use-value framework,' and realised even as my pen was writing the words that my imagination, and my own nature, never think of nature in this way. I never think what *value* has the countryside? And, forced to now, cannot compute it in actual or monetary terms. Which is maybe why I am not a real farmer, and will never threaten to be wealthy. When I think of the countryside, I think of it mostly in spiritual terms. Now I know this is a tricky one. The foundations of my thinking probably come not first from life experience but from literature. They come from Yeats, and Synge, who, maybe because he came from Dublin to the West, became a kind of literary signpost. They come from our first days, literally landed in Kiltumper, entered me then through my senses, and into what I suppose I consider my soul.

But what the purpose of the countryside is in the twenty-first century has come into keener focus for us here this year. This year many turbines have appeared on the horizon, by day 'jerkily moving armies of gesticulating giants', to quote the late, superb cartographer and writer, Tim Robinson, and, by night, a cordon of red lights against the night sky, an air-fence, a colonisation of the dark. Like us, Tim could not denigrate the word 'farming' and instead called what the turbines are doing 'wind-mining'. He decried how they impose sameness on a varied landscape, how they meant locked gates and metal fences, and he asked that difficult question: 'How much of the world do we have to spoil in order to save it?'

Which gets to the essence of the thing that I have been twisting on all year. The world needs to abandon fossil fuels, yes; we need fully to embrace renewable sources of energy, yes. And is that the end of all arguments so? Do we race ahead and fill the countryside with wind turbines? Is this the purpose of wild and remote places? To be employed in the cause of keeping the lights on in the urban ones? This seems less about changing the way we are living, taking responsibility for resources and how we use them, and more about simply changing where the energy is to come from. Replace the fossil fuel with wind and sun and the cities of the world can carry on as before, illumined through the night. It seems to me a caustic irony that in the rush to embrace a greener way it is the actual green places that count least.

I am well aware that in all of this I am neither expert nor statistician. I am only speaking, literally, from ground level, and from a personal wound in seeing a landscape I have loved

252

for more than half my life unalterably changed by the arrival of steel turbines. I am aware too that if these pages are found and read fifty, sixty years from now, they may seem a paean of naivety – *What was he thinking? Did he not realise what was to come?* And yes, perhaps it is inevitable. Perhaps, by the end of the twenty-first century many, many more wild places, like ours deemed 'not of scenic value', will be filled with turbines, the grass covered with solar panels. ('Scenic value' is, of course, planner-speak, and misses entirely the actual value of landscape, which I would argue is not quantifiable, because that value is and always has been spiritual, is, like spirit, unbounded, and feeds an essential need in human beings for which there is no substitute. But I won't keep going on about this lest, to paraphrase Seamus Heaney, I have the veins in my biro bulging. Enough to say, one landscape cannot be adjudged of value against another. *Please.* Each green and natural place is its own living expression.)

However, mankind being mankind, it is a given that energy use will each year not decline but increase, and more and more sources will be required, more and more land needed. So, as the cities continue to grow, who will say, 'We can't exploit the countryside any more?' Who will say the human and spiritual value of the countryside is too great to be spent simply as a resource for energy?

The gloom of this thought can undo me, the idea that this year is only a fringe-time of what is to come in places like this, and that in places like Kiltumper the future will be much less green for going Green.

*

And then comes the day we have marked on the calendar six months ago just with the word 'Galway'. We know what that means. Chris has already had the blood tests and the scans, and mid-morning we drive out of the townland with the three printed sheets of the results. There is a special kind of aloneness and bondedness too in this. You are in the traffic of the everyday, you are just another car driving to Galway. You offer the ordinary small salutes and waves to those you pass going slowly through the village. Passing the church, I have a silent prayer-reflex, issuing from somewhere deep in me, but I say nothing. By chance, Father Peter is crossing from the church to the shops. He raises a hand in salute and I stop for a moment in the middle of the road and he comes to the window. 'Good luck today,' he says, without my telling him where we are going. Through the years of the cancer he had called to the house many times, and we were always grateful. 'Thank you, Peter,' Chris says. There is just an exchanged nod, then he steps back and we drive on, in a way I can't explain, the better for it.

We hope for – we don't go so far as to dare 'expect' – good news, but we don't know, and the journey is tight with tension and things unsaid. We have made this journey to Oncology at the University Hospital so many times now that it is sort of imprinted on our spirits. We know the routine and the rhythms of it; we know it, like second nature at this stage. It has the familiarity of a recurring dream, you are inside it and can do nothing but follow along. All is the same until we go down the corridor and into Dr Leonard's office.

'I get to see you again today!' Chris says, with a tentative smile.

'Yes,' he says, smiling back but indicating nothing. He turns quickly to the computer screen where I can see the tabulation of figures and test results, but not make out what they mean. He moves his finger down the screen looking for one in particular. Then he turns to Chris's fat paper file on his desk, flicks some pages, 'Do we not have…? Yes, here it is.' And he looks at whatever that was, and then back to the computer. He clicks onto a new page, more numbers, back to the paper file again. All the air seems to have left the room. Behind him, the blind on the window has been pushed crookedly to one side and the view is unremarkable, the backs of buildings, a full car park, a line of traffic trying to get somewhere else. Unlovely and ordinary. So many times over the past few years we have been sitting in this situation, wordless and tense, waiting for the report to be scrutinised, and then hearing: 'There's just a few things we need to keep an eye on.' And Chris has come away with the sense of not being free of the cancer, not able to draw a line under it, and move on into the next part of her life.

So, this time we sit, waiting, saying nothing.

'Your results are good,' says Dr Leonard at last. He seemed to say it so softly, and to the page he is looking at in the thick file, that I am not sure I heard him right. He puts down the page and looks one last time at the screen of his computer.

'They're good?' Chris leans forward to ask.

He scans and scrolls, waits, looks, then, at last, turns to her. 'Yes, yes. Well, they are excellent.'

She wants him to say it again, as if the fact of it is somewhere not yet declared and true. 'Excellent? Really?'

'Yes.'

'Really?'

'Yes.'

For a moment I can imagine this exchange continuing on into the night, 'Tell me again.' 'Excellent.' 'One more time, please.' As though the telling will take the same five years of living the illness took, but I cut it short by turning to Chris and we embrace awkwardly in our seats.

'So, is that it, then?' she asks.

'I think so. Yes.'

'I'm done with Oncology?'

'Yes.'

'I don't get to see you any more?'

'Well. No, not unless you need to. We're always here. Call us anytime if you're concerned. You're all done here. Congratulations!'

There is a strange conflux of emotions in this. There is the sense of an ending, only a good one. This is the last in the long line of days that began when Chris was struck by that severe pain in her gut in London, a long line that has led at last back to this moment. To *That's it then*.

But there is also a strange, completely natural, fear in being set free.

You've been closely monitored for years, and now? It takes some mental adjustment to step away from the identity of oncology patient, to trust all will be well now, and it can't happen in an instant.

But here's the first step.

We stand up. Chris gives Dr Leonard a copy of *This is Happiness* and thanks him. There are tears close, but they don't fall. We shake his hand and then are back in the corridor where the next patient is waiting to come in. There is too much to say to say anything. We hug each other. I am flooded with a gratitude that could be translated as a sense of blessing, and all those days when I thought Chris would not make it, they rise and float away from us now. We go back down the two flights of stairs and out the doors into the ordinary day once more.

Some days, although not as often enough as I might, I walk the side road, the boreen up to the Blessed Well. (I walk this way more often now, when I don't want to pass the turbines.) Down through the years we have called it Sean's Road because of the two brothers, Sean and Ralph, who lived where the road ends. When Sean died, it became known as Ralph's Road. Then Ralph died and the farm was taken over by the Downes, but for some reason we never called it Downes' Road. We call it The Road to the Blessed Well, now. The council doesn't take care of it, bordered by all sorts of bushes like snowberry (Symphoricarpos albus) and sally bushes, brambles and purple loosestrife and giant hogweed and nettles – now N, this is what I call a ragged hedgerow.

This afternoon I passed through the Downes' farm gate towards the Blessed Well and just where the road turns there is a very ancient and very large evergreen. A Lawson cypress. There's another one arching over and above the lovely old Blessed Well. A hundred or more years ago, they must have been planted as a pair even though they are fifty feet apart. There was probably an old right-of-way along the stone wall that separates the two farms and the trees were landmarks to the well. As I was passing I noticed, for the first time in all the years of walking there, that there were a few saplings growing a few metres from its base. The saplings came away easily.

I forecast the tree will break in half soon. Already this winter, we seem to be in an extended cycle of wild, storm-force Atlantic winds. I arrive at the Blessed Well, the place where I had spread the ashes of my brother Stephen, the place where our daughter got married, the place under the old cypress tree where a small altar is set inside a stone dome, where inside the dome a statue of the Blessed Virgin is decaying. She is surrounded by spent tea lights and empty jars where wilted wild flowers are brown and grey until replaced and refreshed by parishioners stopping by. From my pocket I take three tea lights and place them inside the rose-coloured glass candle jars. Using myself as a shield I strike several matches until they are lit. Stepping back, I say a prayer and give thanks for landing on the right side of the oncology statistics.

I brought the saplings home and N planted three of them in the glasshouse (while we decide where to put them permanently so I don't plant them in the wrong place!). I put one in a vase, brightening up the back bedroom where in a few weeks one of our children will sleep – come home to celebrate Christmas.

Christmas morning. At eight o'clock, a rosy light in the eastern sky on the far side of the Grove. Silently I watch it spread. Watching it, you have the sense that it is a completed image, a kind of Supreme Painting, all splayed pinks and whites and greys laid on the broad canvas of the sky above the bare trees. Your eye takes it in as one picture, Christmas Dawn, and because it is Christmas morning and holds within it all sixty of the ones I have known, they are all in some lived way present. I don't mean in exact or remembered detail, I mean the profound familiarity of this Christmas morning stillness before the day proper begins. There is a parcelled hush to it that I recognise as something I deeply love. All is prepared, and there is this poised moment, for which I feel a kind of reverence and gratitude. The light is coming, is what I tell myself. The light is coming, is my silent prayer.

And soon enough the rosy sky that I thought of as a completed painting is already changed and changing, the colours steadily stretching, paling and dissolving as the light comes towards us. It only takes minutes, but I will keep it with me the rest of the day.

There are three pheasants on the steps by the back door, one female, two males. I watch them through the glass, pecking and foostering in the grass for things invisible. When I open the door as gently as possible to throw out some crusts, the three birds startle and with clatter and squawk the two males take off, leaving the female some moments as her eye darts in neck-twists from the crusts to the now-flown gentlemen. There's a good thirty seconds while she makes the calculation. Not to be a factor, I stay perfectly still, and in the end, she chooses the company over the bread and shoots off across the grass, making a good business of complaint in her cries to the two un-brave boys. They all re-meet in the safe harbour of the bare sycamore that has been three times hit by lightning. She carries on making a good squawking, giving out, I expect, and not feeling too much of the Christmas spirit.

I leave the crusts where they are; it's between them and the magpies.

Well, whichever, Happy Crustmas.

I go around the front of the house. All the branches of the still uncut-down dead cherry tree wear a frilled, sage lichen. We have no outdoor Christmas lights in the garden, but in the morning light today by nature the tree looks decorated, and some inner voice says, *You see?*

And I do.

Not everything is grown for its bloom, and decay can have its own grace I tell myself.

Remember that.

I will try.

At my feet, Thanks, looking up at me for food.

It is an extraordinarily still and mild early morning. The air doesn't move and the whole townland has a sense of suspension. There's something tender in it, the stillness in the landscape and the activity I can imagine inside the houses. I have a momentary impression of being outside of life, a witness, almost exactly like that captured in *It's a Wonderful Life*. There is, in my case anyway, also a feeling of what I believe is holiness. It is imposed, I know, the day like any other in reality, but the imprint of millennia is not to be easily ignored, and there is for me a spirit-quality in standing in the green hush this Christmas morning and feeling in the Kiltumper garden something bestowed.

*

This year the *Sunday Times* ran an interview with me under the headline 'Happily Out of Step' and while I couldn't read the article – have never been able to read anything about me or my work (in fact can't really think of my writing as 'work' – does a chestnut tree think of chestnuts as 'work'?) – I couldn't avoid the headline. And for a good few days after, it came for me, the out-of-step bit, and it stuck, because of course there was some truth in it.

I am, and always have been, out of step. Whenever I have ostensibly been 'in step', working as a teacher, say, or going into church with the rest of the congregation, there I have felt I have been performing. (In another interview this year I was asked if I suffered from imposter syndrome as a writer – no was the answer. Never as a writer; as a human being, yes, sometimes.)

So, I have been thinking of what it means to be living an out-of-step life.

Coincidentally, if you believe in coincidence, I have been reading Thomas Merton. And in his *Raids on the Unspeakable* there is this: 'There is no explanation and no justification for the solitary life, since it is without a law. To be a contemplative is therefore to be an outlaw.'

And I think I agree that the out-of-step-ness comes with a sense of outlaw, although in my case without the heroic connotations. I don't believe I had a choice, that if I was to be true to myself and my nature I had to live and think and be this way, and this brings me to the bigger picture of the way I – and more importantly, we – have been living life. Chris too is out of step, only in a more beautiful and natural way than me. Her feeling for nature, her *personal* response to what is growing, to beauty and nature in general, are not within the realm of the 'normal', or of those in step. Chris is more sensitive to nature than anyone I have ever met, and 'more' of anything places you on the outside.

But this year, I realise I have begun not only to reconcile to this, but to embrace it. And that is one of the acts of these pages. Slowly, a kind of contentment has grown in me, which I recognise as coming from staying true. But also, today, I am gifted a little general hope for the world, and I don't brush it aside. I think that the action of the year just gone has been to move the world towards a different step. The year, it seems to me, has been notable for rising awareness of the needs of the planet, and while some of this is only at a general and vague level, and it is clear that quick-fix, knee-jerk responses are not what is called for, the urgency and outcry are real, and hopeful.

Now, here, the action of the year, the road-widening and the tree-felling, the erection of the turbines behind us, has made our home place in Kiltumper feel more out of place. The garden feels like an anomaly, albeit a beautiful one, and almost, because much more exposed, on an edge of sorts. Ours is like no other house or garden here, is in fact blatantly (*bloomingly?*) out of step. Of course, looked at differently, on a deeper level perhaps, more in step.

Well, turning over these thoughts today, I came to a sense that who we are, what we are, and *where* we are, all sit at ease, organically, harmonically, like all those plants I can see outside the twelve windows of the conservatory as I am writing this.

We are this place, in this moment.

The rain comes down soundless and soft, and two birds I am watching seek shelter in the pale grey of the weeping pear, and my life, the writing, the living, the loving, my truest friend in this life, Chris, the children, the garden, the house, the whole entirety seems momentarily at one, seamlessly in step in fact. And, for this moment at least, it all feels, well, right. And as I sit here, there is a profound peace in that, and I am filled with an immense gratitude that is another word for love.

For Chris, for the children, for this place, and our being in it.

I will go out the door now into the softest of rain.

There is sure to be work to be done.

Papaver Orientale

13

Summer 2020

The future when it comes is never the one imagined. And this year the first thing to say is, simply, we are well. Our children, family, friends, in their distant places, today, are well. And that is everything. They are healthy. Perspective is only the view from where you are standing, and in the green isolation in Kiltumper today our view is all garden, the people all invisible in various elsewheres. We have the same loneliness that everyone else has now, the same loss that is sealed in the term 'lockdown', but we have the garden, and while it was always the case, perhaps now in a more obvious way, it is clear it has been our saviour. It is the place to bring your upset, your grief, your aloneness and anxiety. It is the place where you can believe in a future, which these days, in maybe the mid-point of a global pandemic, is not so certain. But a seed has a future. A seedling does, a plant does, and in your participation in that, in its smallest moment, is hope. This, I think, must be some part of why, for maybe the first time in recorded history, there were queues outside the garden centres waiting for them to reopen. Yes, people were worried about the food chain, wanted to grow fruits and vegetables, but not only that. I think there was a deeper impulse, to do with time and cycles and the future. This is something I often come back to when daydreaming in the garden. How a garden is always in the now, and yet seeded in your every action is a future, a best version of the plant, the bed, the imagined next spring, next summer, that you place your trust in, in part because we live and garden in a round, and *because it has always been so.* And today, five months into the pandemic, with no clear sense of its end, what the rest of this year, or what next year will be, the consolation of that seems more potent. In so many ways the world seems less firm than it was. Which may part-explain why so many have turned to the firm ground of gardens, why, whether back yard or pot-on-a-windowsill, growing something is what so many of us lean towards. A seed put in soil is an act of faith, action of faith too, for it both imagines and makes happen a better tomorrow. And that is what Chris and I cling to, as we go quietly into the garden each day.

*

At the start of the year the turbines began to turn. Despite all my attempts over the previous year to talk away the threat of them, to tell Chris they would not be so loud, that we could learn to ignore them, once they start the blades blow away all talk and the reality shocks both of us. I get a sickening feeling in my stomach. When the wind is from the west, the whoop of the spinning is louder than I thought possible. If we open the kitchen or bedroom windows, we can hear it. If we go out the back door or walk towards the glasshouse or tunnel it is the first thing we notice. The motion of the blades catches your eye. The chopping sound is inescapable, the air polluted with the constant mechanical noise.

'We will get used to it,' I say to Chris. But my heart is not in it, and I am not sure I believe myself.

It is everything Chris hoped it would not be. 'It's like an echo chamber inside the polytunnel, Niall. I can't work in there today.'

But the wind is not always from the west, and so while the noise is always there, it is not bad every day. We learn wind direction by ear. A good day is a still day, *Thank you, God,* or a day when the turbine nearest to us is broken down and stopped, sitting cruciform on the top of Tumper, which gladdens Chris's heart, and ears. The dealing with this is literally isolating, you are islanded by what your eyes and ears have to deal with and, aware that this is a counter-narrative to the mainstream of green energy stories; not infrequently do we feel we are a Mr and Mrs Quixote defending a position already lost.

One evening, a man who sometimes walks his dog along our road stops at the gate and says, 'Can I just ask. How do you cope with it? I live a mile and a half away and sometimes I can hear them. How do you live with it? And how is it allowed?'

Eight months after the contractors leave, the road remains unrepaired, the stone walls and ditches not remade, orange traffic cones blown over, and after all the lorries and diggers and deliveries, the road has an abandoned air, the erectors gone to raise turbines in the next green place.

*

Today, thinking about John Donne's line about one room being an everywhere, and wondering: if a room can be an everywhere, what then might a garden be?

With cafés and restaurants closed, Chris and I arrange to meet for coffee in the lower end of the garden. I move the two Adirondack chairs our children gave us for my sixtieth down there. We have never sat and seen the garden from this perspective. The sun shines down, the birds sing.

'Nice café,' Chris says. 'But I can see the top of the turbine spinning.'

I am already looking for where tomorrow's café will be. New location, same crowd.

*

April is an April out of a gardener's fantasy. The days are mild, even warm. The plants in the garden rise unhindered by the winds and rains that normally come. I imagine at first a caution in them, in the frail first emergence out of the dreamtime of winter. They know they are in Kiltumper, they must know the weather as well as the soil (*and the gardeners?*), know what to expect, so this April is as surprising to them as to us. And a week becomes two becomes three, luminous sunlight, day after day. There is no wind. We eat in the garden, live in the garden, and each day watch it become its best self.

That because of the lockdown no one can visit, no one but the two of us can see this garden in its April glory, is somehow moving to me. There's a pathos to the idea of all the gardeners everywhere in gardens that cannot be visited, but still doing the tending. The small worlds of plots, allotments, where this astonishing and glorious springtime is rising. But also, a dichotomy, as each evening we go inside and hear the latest figures of the virus, the larger, stricken world outside the hedge where those on the frontlines are every day in danger. Both are real, the magnificent spring and the appalling disease. 'The marvellous and the actual', I remember, is Seamus Heaney's phrase. The duality of our lives' experience seems more pressing these days when the greater reality is almost overwhelming and the local marvellous must be held on to when it comes: three dozen pink tulips Chris has grown for our son's wedding celebration, which has been postponed. They are flawless. The twenty white-and-apricot foxgloves she set from seed last autumn, potting them along into ever larger pots for table decorations. The wingbeat of that bird overhead as I write this. Pause. Listen. Yes, this is a time to get through, but it is also *this time*, which, as they say, will not come again. As if by clockwork, the cuckoo returns on 21 April, soon after, the swallows do. These things take on an added force; during a time of so much illness and dying they are a pulse-beat that says we are all still within a cycle of life. I don't think we have ever been so glad to see the swallows.

'Welcome,' we say to their flicker-flight up in the rafters of the open cabin. 'Welcome.'

If April was like none before, May is too. Days of perfect sunshine. We are in the garden seven days a week, and like all relationships it benefits from the time given to it. *If a garden could smile*, I write in my notebook. That this is therapeutic for both of us we don't need to say. It is a given, in every sense. In the uncertainty over everything beyond the hedge, the poppies, the peas, the tomatoes can be counted on.

And in a real way we do count on them. Not just for the food – although this is important, and by midsummer we are eating from the garden every day (lettuce, runner beans, peas, a new cut-and-come-again sprouting broccoli, kale, three varieties) – not just for the beauty – also important to our living here – but for the continuance. For the carrying on, for the simple fact of being part of, participant in and witness to, a cycle. This, for me, might be the most important part, for what it feeds is your spirit. And as the pandemic goes on, this has a value that cannot be measured.

*

These days the place reasserts its own primacy over people, that is, its life is the life of plants and its sounds are the sounds not of people but of birds. It is universal, there are articles, commentaries, news stories in many countries about birds. Are there *more* this year? Or are we more keenly witnesses? I remind myself to look up Lear's image of God's spies. Birds, birds everywhere.

*

'Niall, Tim's here!'

Now that we are in the time of going nowhere and seeing nobody, the postman's van regains its importance of long ago, and we are on the watch-out for it. Tim pulls up at the gate and one or both of us rush out to meet him. Many days he is the only other human being we see, and from him, standing at our two metres distance, we get the news of the greater world, by which I mean the parish. It's hard to quantify what this means in a rural place, it's unquantifiable to government policymakers, but it's a lot, and in human need feels more significant than anything virtual. The talk might be brief, might be only the weather, the unrepaired state of the road or the noise of the turbines while we are standing there, but there's something vital in it, and we are grateful.

Knowing that Chris is considered to be in the new category of the vulnerable, and that we are hardly going outside the hedge line, Tim reaches for his pen and on a scrap of paper writes his mobile number.

'If you need anything brought from the village,' he says, 'just call.'

There's nothing to say but thank you, and feel in this small instance the reality of what it is to live in a community.

And throughout the parish there are many examples of the same, volunteers and support systems, and my sense is of a connective tissue of care, invisible, understated, but there all the same, and going some way towards proving the universal truth that desperate times bring out the best in people.

***

There are moments in gardens you can't script. And with Niall these can happen anytime. He is so earnest. Once, in a surge of inspiration he dug up some of our potatoes and made a lovely potato salad. While I was eating it, I thought it tasted a bit like perfume. 'What did you put into the salad?'

'Rosemary,' he said, proud as punch.

A gardener knows the taste of rosemary. When we were out later, I asked him to show me which rosemary he'd used.

'Ah, I see. You know, I would never have thought of adding lavender to potato salad.'

Lately, with all meals being homecooked, he has been doing more cooking than usual, and is becoming an excellent cook. If I haven't already brought it in, he asks what's ready for harvest this evening, and goes out and gathers it.

'The new kale. Not the curly or the black.'

'I know, I helped plant it.'

I went back to writing something I've been working on while he cooked dinner.

And I have to hand it to him, the kale was delicious.

But the next day, as we were doing my favourite thing in the world – working side by side with Niall in the garden – I noticed some plants inside the tunnel were rather bare, and missing some of their leaves.

'O dear. What happened to these?'

'We had them for dinner last night,' he said. 'The kale.'

It took me a moment. 'Niall! Those are not kale, they are the newly planted purple-sprouting broccoli.'

'Oops.' He made the face of apology, then he grinned. 'Well, new discovery, turns out some of the most delicious kale on the planet is the not-purple leaves of purple-sprouting broccoli.'

Although we have several large water butts, and after the previous droughts use the water sparingly in the summer, by June our well goes dry again. There has been no significant rain for seven weeks, and the water table has sunk somewhere beyond reach. Whether this is the future for summers in the west of Ireland, droughts and downpours, or not, is hard to say. A Christmas gift from Martin and Pauline of a large rectangular clear plastic container that holds a few hundred litres of water sits now empty behind the polytunnel. But, in a further proof that the spirit of community we found here thirty-four years ago still endures, unbidden, Michael down the road arrives in his tractor with a creamery tank filled with water. In his seventies, he manoeuvres himself down out of the cab with a boyish grin. 'Don't be short of water,' he says as we fill the container to the brim.

As he's leaving, he pushes open the door to ask, 'When she's passing, do you think Chrissy could take a look at my tomatoes?'

'Course she will, Michael.'

She calls in the next morning.

Somewhere between too little and too much watering the perfect tomato lies, and it is a strangely comforting thought that because the conditions of each summer and each location are different, no gardener can really say exactly how much watering that is. So, it seems to me, the talk of tomatoes between two growers is always the same, always about '*this year's*', in which all past and all future ones count for nothing, this year's are the only ones that matter, and how to read what the plants are saying mid-season is a good portion of success or failure.

'That one's a goner, Chrissy,' Michael says, tossing his head towards a particularly shrivelled plant.

'It is, Michael.' She pulls it out of the ground for examination. There's a moment for the obsequies, then: 'I have a spare plant I'll bring down to you.'

'Lovely, Chrissy.'

Two weeks later she brings him a cucumber plant. 'You like cucumbers, Michael?'

'I do, Chrissy.'

For five weeks we have no water in the house. The bottom of the well is wet mud, then dry mud, caked and cracked, and the skies are blue. Michael comes with the creamery tank every three days. We go to neighbours for showers and to fill drinking water, making sure we adhere to the phrase of the year, social distance, and it becomes a routine and rhythm. Each day Chris adjudicates which plants get some water, and somehow the garden survives.

The anchusa seeds I bought at the Chelsea Flower Show two years ago and planted out in the garden last year have decided this is weather they like; each day I go out they are taller and taller and bluer and bluer. If I'd known they were going to say this is delicious weather, we love it here in Kiltumper, I would have placed them elsewhere to let them have their moment in the sun. They are side-by-side with the purple lupin and a sky-blue delphinium and a baptisia, and none of them has wriggle room.

But this is fine for now, because they are growing so straight and tall and the lupin is sturdy. 'It was worth waiting the two years for this,' Niall says, looking with some wonder at the blue flowers.

Well, of course it was, it's an anchusa.

⌒≺

I read that in sunshine borage replenishes nectar every two minutes. *Can it be true?* Chris picks some of the blue flowers for the salad today. 'Not lavender, Niall,' she says when I look into the bowl.

Besides the Kiltumper-exotic of eating blue, when I put one of the borage blooms in my mouth, I am momentarily other, thinking of nectar and bees and sunshine, and have, if not a winged feeling, certainly the air-giddiness of eating flowers.

⌒≺

In June, our friend Martin, the children's old headmaster – I don't mean old as in ancient for he's a year younger than Niall – sends me a photo of a tiny white puppy with black eyes and black nose and floppy ears. Beneath it the message: Free to a good home.

We'd been thinking of getting a puppy for two years but couldn't decide upon the breed. When the children were growing up, we had a Golden Retriever named Huckleberry. He lived fifteen years. A gorgeous white Golden that everybody loved, he was famous in the townland not only for his looks but for his good-naturedness, and Martin had taken care of him when we took our trip around the world all those years ago. We couldn't go for another white Golden because it would remind us too much of Huckleberry whose death took us a few years to get over. But Martin's photo makes the decision for us. A puppy half Golden and half Samoyed. Hashtag: Goldensammy. Born in April (the month of all good birthdays in our family) he was five weeks old in the photo. We said yes and then looked at each other. Puppy training!

But if there is ever a time to get a puppy it is now. While we were waiting for him to come up from Cork where Martin's sister Nollaig had a Golden and her husband a Samoyed and eight love puppies the result – we read up about the mixed breed. Nothing but gorgeous and friendly, it seemed, very white and fluffy with two coats. In Siberia they were bred to herd reindeer.

Before his arrival we had several discussions about his name.
'Sammy?' I said.
'No.'
'Goldie?'
'No,' Niall said, a knowing smile breaking across his eyes.
'What? You already know his name?'
'I'll give you three guesses.'
It took me a few seconds. Then I knew too, because Niall is a clever lad and for him everything fits one way or another.
'Ha!' I said. 'Of course. It's Finn.'
'After Huckleberry, has to be Finn.'
And a few days later there he was, a white ball spilling out of Martin's arms into the wonderland of the garden. 'Welcome, Finn.'

___

'I think the future is terrifying,' Chris says.

'It is always uncertain,' is my first response. 'Always unknown.'

But there's no solace in that. Maybe there is no best answer to when somebody says the future is terrifying, and these days when family and friends remain in distant places and the figures of a virulent disease continue to mount daily, it may be absurd to think anything else. One of our children had been laid off early in the pandemic, prospects for work poor, economies everywhere on the edge of collapse. Everywhere you look it *is* frightening. And so, I give the only honest reply, 'I know.' After a little time, Chris says what she always says in low moments, 'I'm going into the garden.'

'All right.'

She is silently, almost meditatively, deadheading when I go out to join her. There is no discussion. I gather the fallen flowers as she goes. The evening comes around us.

*

'We need to tie everything up,' Chris announces as Storm Ellen approaches. We begin with stakes and a roll of green windbreak, which we set up around the new copper beech hedge that one day will part-screen the kitchen window from the turbines. When we are done, the hedge is inside a rectangular green room, and without saying so, I know we both feel *That is something.* This part of gardening, the *Something done today*, seems more important than ever, and not for the first time I am grateful that we have a garden. Caring for it is purpose enough. Silently we work our way through the beds, Chris, stepping gingerly in among the full and rich growth, her face lost in a silver fall of hair as she leans down and tries to find the place which will not damage the roots, me passing in the string, pressing down the placed canes.

The anchusas are taller than I am. I am on a chair, which is a bit crazy, to hammer in the canes and wind the string around them. When finished, it looks like a spider has woven a kooky pattern through and around, criss-crossing the patch of blue and purple and green. The anchusa is no longer its magnificent self, standing alone. These days I am often on the point of giving up. But maybe it will survive.

It's maybe a thing with all gardeners as the summer runs away and a storm is coming. Part of me thinks I should just let the garden go, let the chips fall as they may. But then I look at all the plants and think: how can I not protect this one? And then that one?

⤝⤞

A storm in summer can be disastrous for the garden; the full leaves, the heavy blooms, cannot prepare for it, nor can they quite be protected. 'This dahlia won't survive,' Chris says, taking the secateurs to three enormous cerise blossoms. They are too heavy for their stems. On a fine day you can look at them and think the mechanics, the balance between flower and stalk is mysterious and sublime. But the gusts and downpours that are coming will take them to the ground. 'Better you come inside,' she says.

Having gone through the beds staking everything once more, including the various giant cosmos that grew this year, we shut the windows of the glasshouse and tie the doors closed, carry to the cabin the garden furniture, take down the geraniums from the table, shut up the polytunnel. All of this is routine now, familiar to gardeners in Atlantic places, but today we do it in the shadow of the long talk about the pandemic, how in the spring it was imagined that by August it would be, if not gone, at least controlled. And jobs would return to the many who lost them. But today that is not the case. And so, yes, the future is deeply uncertain.

The storm is arriving as we are finishing. It comes into the garden from the south and surprises the giant berberis that is used to taking the western storms on its back, which at this stage is bared to thorn-wood. The wind is fierce, un-foots Pythagoras's pea fence, tilts all canes and plants with them, and makes a sudden rain of green leaves. The south side of trees and plants are stripped. Leaf-swirls dance on the pebbles, rise and twirl and come back down again. We are tempted to watch the front beds from the window, but the rain and darkness come together like a curtain, and like a curtain coming down we think, *That is the garden for this year.*

*

I am out early in the morning with Finn. There is the stillness that people call unearthly. The garden is bright-littered with pieces of itself. Everywhere are flecks of colour, the red of new fuchsia blossoms, the pale purple of the phlox and pink petals of Japanese anemones and dahlias, and when your eye spots one automatically you look back to seek the plant it came from. And are surprised to see: it is still standing.

Chris comes outside with her mug of tea, and, the same as every morning, as if he has not seen her in an age, Finn charges to her. She bends down to him giving him the hug he requires, and returns, while her eye is taking in the garden. After the storm there are broken stems, the baptisia blown in to mix with the anchusa. 'It's all looking a bit like a dessert gone wrong,' Chris says.

'But not so bad,' I say, playing my part.

Already, in an instinct old as humankind, she is righting the anchusa, moving to straighten the canes around the reflowering delphinium. Then I am straightening the ones around the lupins. Finn is in thrall to a fallen red dahlia bloom and finds now he can throw it in the air.

As we go from plant to plant, with Finn in tow, I have a deep calm that I can not explain. *This matters*, this simplest of acts, checking the garden after the storm, tending it. In some way larger than we ourselves, it matters. And what comes to me then is the sense of all the other gardeners, in all the gardens large and small, at this particular time in the world, also working, bending and unbending, like us, wordlessly moving from one plant to the next. And here in the garden in Kiltumper the image of that, the oneness of all of us, gardeners, everywhere, tending to plant and soil, fills me with a real and tangible hope.

'Not so bad,' I say again to Chris.

I don't mention the storm, or the fear of yesterday, or the greater one of the future. I don't need to, she knows.

She straightens up alongside the baptisia, takes a measured look out across the beds, then turns.

'No, not so bad,' she says. 'Not so bad, is it Finn?'

THE END

Acknowledgements

No garden is made without the help of many people.

From the moment we arrived in Kiltumper we have been supported by many. This continues today. To list them all would take a book in itself. The early garden was created by the late Mary Breen and her late brother-in-law Joeso, and we will always be grateful to them for the welcome and advice they gave us. We would like to particularly thank our wonderful neighbours Martin and Pauline Hehir, all the Downes family, Colette and Jack Mannion, Bob Choppin, Tom Keane and our great friend Martin Keane, and all who encouraged us to stay the course, especially our families.

Likewise, no book exists without the encouragement and support of many. We would like to thank Michael Fishwick and all the team at Bloomsbury, especially Sarah Ruddick; Grace McNamee and all at Bloomsbury USA; our indefatigable agent Caroline Michel and all at Peters Fraser & Dunlop.

And finally, this book is for our children, Deirdre and Joseph, around whose childhoods the garden grew.

A Note on the Authors

Niall Williams was born in Dublin in 1958. He is the author of the Man Booker-longlisted *History of the Rain* and eight other novels including *This is Happiness* and *Four Letters of Love*, set to be a major motion picture.

niallwilliams.com

Christine Breen was born in New York, and educated in Boston and Dublin. She is an artist and the author of the novel *Her Name is Rose* and a travel memoir, *So Many Miles to Paradise*.

christinebreen.info

Together, they have co-authored four non-fiction books about their early life in Kiltumper. They have two children, a cat, a dog and a very large garden in County Clare.

A Note on the Type

The text of this book is set Adobe Garamond. It is one of several versions of Garamond based on the designs of Claude Garamond. It is thought that Garamond based his font on Bembo, cut in 1495 by Francesco Griffo in collaboration with the Italian printer Aldus Manutius. Garamond types were first used in books printed in Paris around 1532. Many of the present-day versions of this type are based on the *Typi Academiae* of Jean Jannon cut in Sedan in 1615.

Claude Garamond was born in Paris in 1480. He learned how to cut type from his father and by the age of fifteen he was able to fashion steel punches the size of a pica with great precision. At the age of sixty he was commissioned by King Francis I to design a Greek alphabet, and for this he was given the honourable title of royal type founder. He died in 1561.